RECOMBINANT DNA METHODOLOGY

RECOMBINANT DNA METHODOLOGY

Editors:

Jo-Anne R. Dillon
Laboratory Centre for Disease Control, Ottawa

Anwar Nasim
National Research Council of Canada, Ottawa

Earle R. Nestmann
Department of National Health and Welfare, Ottawa

JOHN WILEY & SONS

New York
Chichester
Brisbane
Toronto
Singapore

Library of Congress Cataloging in Publication Data:

Dillon, Jo-Anne R.
    Recombinant DNA methodology.

    Includes index.
    1. Recombinant DNA.   2. Genetic engineering—
Methodology.   I. Nasim, A. (Anwar), 1935–
II. Nestmann, E. R.   III. Title.   IV. Title: Recombinant
D.N.A. methodology. [DNLM: 1. DNA, Recombinant—
laboratory manuals. QU 25 D579r]

QH442.D55   1985   574.87'3282   84-20946
ISBN 0-471-89851-1

Printed in the United States of America

10 9 8 7 6 5 4 3 2 1

PREFACE

This laboratory manual is intended as an introductory text for students and researchers new to the area of recombinant DNA technology. Our interest in compiling this handbook was stimulated by several laboratory courses on recombinant DNA methodology that were held in Ottawa over the past three years. The course participants included senior graduate students and professional scientists from a variety of backgrounds and institutions. They were all united, however, by the desire to apply the techniques they would learn to their different research projects. The initial framework for this text was constructed from these courses.

The basic techniques, outlined in Chapter 1, form the cornerstone on which more advanced or varied techniques may be built. Other chapters provide basic information on methods for cloning in yeast and *Pseudomonas aeruginosa* hosts, and the final chapter outlines a variety of methods for site-directed mutagenesis. Throughout the book, an attempt has been made to give the reader pertinent, but not exhaustive, background information on each technique. Where possible, techniques have been generalized to encompass conditions for a variety of microorganisms, rather than limiting examples to a particular phage or bacterium. Also, alternative methods ("options") for many procedures have been included to encourage workers to adapt and modify the techniques to their own purposes. The format used for the methods was chosen so that the reader could understand the purpose and pitfalls of each step in a given procedure. After introducing the techniques and providing brief background information, the steps involved in carrying out the procedures are detailed. Comments explaining why certain operations are performed, as well as details of reagent formulations and alternative techniques, are presented in parallel. A number of readers may be unfamiliar with some of the terminology, so an introduction to important terms is included at the front of the book. For the readers' convenience, the appendixes provide a summary of the formulae of the buffers, reagents, and media, and detailed lists of companies providing chemicals and equipment. The index has been compiled in such a manner as to cross-reference both the terms and techniques used throughout the text, as well as the composition of buffers and other reagents.

This manual is not intended to be an exhaustive compendium of all recombinant DNA techniques and host-vector systems. Thus, several topics, particularly those in which the authors have had no direct experience, or that have been excellently covered in other books, have been omitted. Such topics include eukaryotic host-vector systems, and the development and use of specific vectors (for example, λ).

Most of the contributing authors for this manual also participated in teaching the recombinant DNA courses described above. Therefore, their contributions, both in teaching and to this manual, are deeply appreciated. In addition we thank the students from the courses for providing useful comments concerning the content and format of this manual. We wish to thank Douglas Johnson (University of Ottawa) and David Thomas (National Research Council, Ottawa) for reviewing portions of this book. We also thank Greg Bezanson and Him Yeung for their help in compiling the introduction to important terms and the appendices. Finally, we wish to thank Mrs. Lynn Anderson who has been actively involved in preparing course manuals and in typing this text. Her stylistic and editorial suggestions, as well as her patience, have been invaluable in the preparation of this manual.

Jo-Anne R. Dillon
Anwar Nasim
Earle R. Nestmann

CONTRIBUTING AUTHORS

Gregory S. Bezanson, Antimicrobials and Molecular Biology Division, Bureau of Microbiology, Laboratory Centre for Disease Control, Ottawa, Ontario KlA OL2 Canada

Jo-Anne R. Dillon, Antimicrobials and Molecular Biology Division, Bureau of Microbiology, Laboratory Centre for Disease Control, Ottawa, Ontario KlA OL2 Canada

Shirley Gillam, Department of Pathology, Faculty of Medicine, University of British Columbia, Vancouver, British Columbia V6T 1W5 Canada

C. P. Hollenberg, Institüt für Mikrobiologie, Universität Dusseldorf, Federal Republic of Germany

A. Nasim, Molecular Genetics Section, Division of Biological Sciences, National Research Council, Ottawa, Ontario KlA OR6 Canada

A. A. Potter, Department of Biology, Carleton University, Ottawa, Ontario KlS 5B6 Canada

Michael Smith, Department of Biochemistry, Faculty of Medicine, University of British Columbia, Vancouver, British Columbia V6T 1W5 Canada

Kwok-Him Yeung, Antimicrobials and Molecular Biology Division, Bureau of Microbiology, Laboratory Centre for Disease Control, Ottawa, Ontario KlA OL2 Canada

R. S. Zitomer, Department of Biological Sciences, State University of New York at Albany, Albany, New York 12222, United States

Mark Zoller, Department of Biochemistry, Faculty of Medicine, University of British Columbia, Vancouver, British Columbia V6T 1W5 Canada

CONTENTS

INTRODUCTION TO IMPORTANT TERMS

ALKALINE PHOSPHATASE

An enzyme that removes 5′-phosphate groups from RNA and DNA. It is used to prevent the recircularization of nonhybrid molecules in gene cloning.

AMPLIFICATION

An increase in the number of copies of a particular nucleic acid fragment resulting from the replication of the vector into which it has been cloned.

An increase in the number of copies of a plasmid effected by the exposure of its host cell to chloramphenicol, an antibiotic which inhibits chromosome replication.

ANNEALING

The process in which two separate strands of nucleic acid interact to form duplex molecules. The interaction is based on hydrogen bond formation between complementary base pairs (adenine-thymine, guanine-cytosine in DNA) present in the individual strands.

AUTORADIOGRAPHY

A method for the localization of radioactive molecules by superimposing a photographic film over a sample. Dark spots on the subsequently developed film indicate the positions of the radioactive material in the original chromatogram or sample.

BACTERIOPHAGE (PHAGE)

A virus that is propagated (grown) in a bacterial cell.

***Bal*31 NUCLEASE**

An enzyme displaying a highly specific single-strand endonuclease activity plus a 3′ and 5′ exonuclease activity that simultaneously digests both strands of a DNA duplex from both ends.

BLOTTING

See Southern Blotting, Colony Hybridization.

BLUNT (FLUSH) ENDS

Termini of duplex DNA that are completely base paired.

CHROMOSOME

The principal genetic element of a cell. It is composed of DNA (and, in eukaryotes, is associated with protein) and carries most of the information necessary for cell survival.

CLONE

A group of genetically identical molecules, cells or organisms asexually descended from a common ancestor. Cloning is the process of propagating such identical cells or organisms. Recombinant DNA techniques make it possible to clone individual genes; this is called molecular cloning.

COHESIVE (STICKY) ENDS

Single-stranded termini of duplex DNA molecules having single strands that are complementary and can therefore be base-paired to other molecules with the same sequence.

COLONY HYBRIDIZATION

A procedure for the direct detection of a particular DNA sequence within an array of bacterial cells (clones) using a radiolabeled DNA or RNA molecule (probe) that is complementary to the sequence being sought.

COMPLEMENTARY (c) DNA

Fragments of DNA that are synthesized on messenger RNA templates (i.e., complementary to mRNA) through the action of a reverse transcriptase.

CONJUGATION

The one-way transfer of genetic material between cells during bacterial mating. The DNA may be either chromosomal or plasmid in origin. A conjugative plasmid can determine the transfer of its own DNA to another cell. It is self-transmissible. Nonconjugative plasmids are not self-transmissible.

CONTAINMENT

(1) Biological:
The use of genetically altered bacteria, phages, and plasmids that are unable to complete essential functions (e.g., growth, DNA replication, DNA transfer, infection and/or propagation) except under specific laboratory conditions for the purpose of diminishing or eliminating the possibility of their survival and/or transmission outside the laboratory.

(2) Physical:
The use of laboratory design and practice (e.g., limited access, safety cabinets, aerosol control, protective clothing, pipetting aids) for the purpose of reducing the likelihood of personnel infection and/or the spread of recombinant DNA or its propagating organisms.

COSMID

A plasmid containing the *cos* (cohesive terminus) gene of bacteriophage lambda. Cosmids are used as vectors for the cloning of large (approximately 45 kb) fragments of DNA.

COVALENTLY CLOSED CIRCULAR (CCC) DNA

A circle of duplex DNA in which the double strands are covalently continuous (e.g., plasmid DNA). Secondary interactions between separate sections of the circle, via hydrogen bonding, produces a tightly packed, super-coiled molecule. Open-circular DNA is less compact due to the presence of a break (nick) in the continuity of one strand of the duplex.

DIRECT SELECTION

The enrichment of a particular clone or recombinant molecule to the detriment of others in a mixture of forms through the inclusion of a particular chemical (e.g., antibiotic) or the exposure to specific metabolic (e.g., lactose fermentation) or physical (e.g., high temperature) conditions.

DNA POLYMERASE I An enzyme isolated from *Escherichia coli* that displays both $3' \rightarrow 5'$ and $5' \rightarrow 3'$ exonuclease activity in addition to its $5' \rightarrow 3'$ polymerizing activity (assembly of deoxyribonucleotides into DNA). It is used for the *in vitro* labeling of DNA by nick translation.

DNase I An endonuclease recovered from pancreatic tissue that cuts phosphodiester linkages in both single and double-stranded DNA, usually next to a pyrimidine nucleotide, to yield a polynucleotide ending with a $5'$-phosphate group. It is used in combination with DNA polymerase I in nick translation labeling *in vitro*.

ENDONUCLEASE An enzyme (e.g., restriction endonuclease, DNase I) that cuts DNA at sites within the molecule.

EXONUCLEASE An enzyme that digests DNA from the free ends of a linear molecule by the sequential removal of nucleotide bases.

EXPRESSION VECTOR A plasmid or phage cloning-vehicle, specifically constructed so as to achieve the efficient transcription and translation (expression) of the cloned DNA fragment.

EUKARYOTE A higher cell that is characterized by extensive compartmentalization of its essential functions, the presence of well-defined internal structures and the presence of a nucleus composed of DNA and proteins.

FILAMENTOUS PHAGE A sex-specific (male) bacterial virus that carries a single strand of DNA within a filamentous protein coat (e.g., M13, fd). At one point of its life cycle (the replicative form) it becomes double-stranded.

GENETIC COMPLEMENTATION Two mutant genes are said to be complementary to each other if their gene products can be combined to form a functional unit or if the wild phenotype of the organism is obtained.

GENE A segment of a DNA molecule that codes for a functional product such as a polypeptide chain or RNA molecule

GENE FUSION The use of recombinant DNA techniques to join (fuse) two or more genes that are coding for different products in such a manner that they come under the control of the same regulatory systems.

GENE LIBRARY A collection of randomly-cloned fragments that encompass the entire genome of a given species; also referred to as a clone bank or shotgun collection.

GENE MAPPING Determination of the relative sequence and locations of different genes on a DNA molecule (i.e., chromosome, plasmid).

GENE SPLICING The enzymatic manipulations by which one DNA molecule is attached to another. It also refers to the process by which the introns are removed and exons joined during mRNA synthesis.

GENOME The entire DNA of an organism or individual.

HETERODUPLEX A double-stranded DNA molecule in which the two strands originated from different individuals and which therefore do not have completely complementary base sequences.

HOMOPOLYMER TAILING A procedure for joining DNA fragments using the enzyme terminal deoxynucleotidyl transferase to add a homopolymer extension (e.g., dA and dT residues) to the 3'-ends of the fragment.

HYBRIDIZATION See Annealing.

INSERTIONAL INACTIVATION A strategy for the easy detection of cloned DNA by which a foreign DNA molecule is cloned within a functional gene, thus rendering the gene nonfunctional.

KLENOW FRAGMENT A 76,000 dalton polypeptide that is obtained by partial proteolytic digestion of DNA polymerase I. This enzyme possesses the $5' \rightarrow 3'$ polymerase and $3' \rightarrow 5'$ exonuclease activities, but not the $5' \rightarrow 3'$ exonuclease activity of DNA polymerase I.

LIGASE An enzyme that catalyses the formation of a phosphodiester bond at the site of a single-stranded break in duplex DNA.

LINKERS Chemically synthesized oligonucleotides of defined sequence containing the recognition/cleavage site of one or more restriction enzymes.

mRNA (MESSENGER RNA) RNA that serves as a template for a translational product.

MOBILIZATION The process by which the transfer of a nonconjugative plasmid between bacterial cells is facilitated by a second conjugative plasmid.

MODIFICATION A process whereby host cells protect themselves against endogenous restriction endonucleases by methylating certain bases (in prokaryotes).

MUTATION A heritable change in genetic material.

NICK A break in the phosphodiesterase bond in one of the two strands of a DNA molecule.

NUCLEOSIDE A compound consisting of one of the four bases—adenine, guanine, cytosine, and thymine (or uracil, in the case of RNA)—covalently linked to a pentose.

NUCLEOTIDE The basic unit of nucleic acids; a nucleoside that is phosphorylated at one of its pentose hydroxyl groups.

OPERON A genetic unit consisting of one or more related genes that function coordinately under the joint control of an operator and a repressor.

OPERATOR A segment of DNA capable of interacting with a specific repressor to control the expression of an adjacent operon.

PALINDROME A sequence of DNA in which identical base sequences run in opposite directions. Palindromes are often involved in recognition sites for restriction enzymes.

PLASMID An extrachromosomal, covalently-closed circular DNA structure that can replicate autonomously within a bacterial cell and which is normally nonessential to the survival of the cell.

POLYADENYLATION Addition of poly A to the 3'- ends of mRNA in eukaryotes by a mechanism that does not involve transcription.

POLYMERASE Enzymes that catalyze the assembly of ribonucleotides into RNA or of deoxynucleotides into DNA.

PRIMER A structure that serves as a starting point for polymerization.

PROBE (HYBRIDIZATION) DNA or RNA molecules labeled either radioactively or immunologically which are used in hybridization experiments to detect the presence of complementary sequences.

PROKARYOTE A simple unicellular organism, such as a bacterium or a blue-green alga, with a single chromosome, no nuclear membrane, and no membrane-based organelles.

PROMOTER A DNA sequence to which RNA polymerase may bind and initiate transcription. The sequence T-A-T-A-A-T-G **(PRIBNOW BOX)** is usually associated with prokaryotic promoters. In eukaryotes, the sequence T-A-T-A **(HOGNESS BOX)** is usually found near the promoter region.

PROTOPLASTS Cells of bacteria, fungi or plants from which the cell wall has been completely removed by chemical or enzymatic (e.g., Glusulase) treatment.

RECOMBINANT DNA A hybrid of DNA molecules of different origin, joined using a variety of biochemical techniques.

RESTRICTION ENDONUCLEASES Enzymes that degrade foreign DNA by recognizing specific sequences internal to the molecule and subsequently cleaving the DNA in both strands at sites either internal or external to the sequence. Host DNA is not cleaved because it has been modified (methylated).

REVERSE TRANSCRIPTASE An enzyme from certain RNA viruses that catalyses the polymerization of DNA on an RNA template (also known as RNA-dependent DNA polymerase).

S1 EXONUCLEASE An enzyme that degrades single-stranded DNA (or RNA).

SHINE-DALGARNO (SD) SEQUENCE The ribosome binding site on mRNA, which in *E. coli* is 3 to 11 nucleotides upstream from the initiation codon, and which is also complementary to the 3'- end of 16s ribosomal RNA.

SEQUENCE ANALYSIS Determination of the sequence of nucleotide bases in a DNA molecule.

SHOTGUN METHOD Cloning of random fragments from a genome with a view to selecting a specific DNA fragment. Usually an entire genome is cloned to constitute a gene library.

SHUTTLE VECTOR Cloning vectors with DNA sequences that permit their replication in both bacterial and eukaryotic hosts.

SITE-DIRECTED MUTAGENESIS Highly specific, predetermined change, introduced into DNA by a variety of techniques.

SOUTHERN BLOT Transfer of separated DNA fragments from electrophoretic gels to membrane filters. The "blotted" DNA fragments are detected by hybridization with DNA or RNA probes.

SPHEROPLAST A microbial cell in which the cell wall has been partially removed by an enzymatic or chemical treatment.

TERMINAL TRANSFERASE	An enzyme catalyzing the addition of deoxynucleotides to the 3' ends of DNA (also called terminal deoxynucleotidyl transferase, calf thymus DNA polymerase).
TEMPERATE PHAGE	A phage (for example, λ) that has the option of existing in either the lytic or lysogenic states. In the lysogenic state the phage, known as a prophage, is integrated into the bacterial chromosome. In the lytic state phage particles, after replication, are released by cell lysis.
TRANSCRIPTION	Synthesis of RNA from a DNA template using RNA polymerase.
TRANSDUCTION	A mechanism of genetic exchange between a host and recipient cell mediated by a bacteriophage that has incorporated host genes into its genome.
TRANSFECTION	Uptake of either plasmid or viral DNA (or RNA) by competent cells (uptake of plasmid DNA is often referred to as transformation).
TRANSFORMATION	Uptake of naked DNA by competent cells, its integration into the chromosome and subsequent expression. (Term has colloquially included uptake of plasmid DNA.)
TRANSLATION	Synthesis of protein(s) from a mRNA template.
TRANSPOSITION	Movement of a DNA sequence from one location to another in the genome.
TRANSPOSON (TRANSPOSABLE ELEMENT)	A mobile genetic element that can change its position within or between cellular genomes.
TRYPTOPHAN OPERON	An operon consisting of five contiguous structural genes that code for enzymes involved in the tryptophan biosynthetic pathway and their associated regulatory elements. The *trp* promoter has been used extensively in the construction of cloning vehicles.
VECTOR	A plasmid or viral DNA molecule into which another DNA molecule can be inserted without disruption of the ability of the molecule to replicate itself.

chapter 1

BASIC TECHNIQUES

Jo-Anne R. Dillon, Gregory S. Bezanson, and Kwok-Him Yeung

I. Rapid Methods for the Isolation of Plasmid DNA

Several procedures have been developed in recent years that simplify the isolation of plasmid DNA. When selecting one procedure over another, such factors as the size of the plasmid(s) to be visualized, the simplicity and reproducibility of the procedure, the bacterial genus to be lysed, the gentleness of the procedure (important if large plasmids are to be detected), the DNA yield, and the subsequent use of the plasmid DNA (i.e., for restriction endonuclease analyses, transformation, etc.) must be considered.

For the rapid isolation of plasmid DNA, one should aim for as few manipulations as possible that will still produce interpretable results. Depending on the ultimate use of these results, pure DNA may be required. The decision as to whether an experiment requires pure or "less than pure" DNA is especially important in laboratories where the cost effectiveness of certain procedures is critical. Screening bacterial colonies for plasmid content, or using plasmid DNA for certain transformation and cloning procedures does not require the preparation of pure plasmid DNA. "Clean" DNA may be required to prepare a photograph for publication, for producing restriction maps, in certain cloning procedures, etc. For these purposes, a series of "options" might be added to the basic plasmid isolation technique. For example, an RNase might be used to eliminate RNA, a phenol extraction step might be added to remove proteins, or the addition of the nuclease inhibitor, diethyl pyrocarbonate (Blattner *et al.*, 1977), might be used to eliminate endogenous nucleases present in such genera as *Clostridium* and *Campylobacter*.

METHOD 1: ALKALINE-DETERGENT METHOD

This method, derived from three groups (Birnboim and Doly, 1979; Kado and Liu, 1981; Casse *et al.*, 1979) is a rapid, reproducible, and gentle procedure for the recovery of bacterial plasmids and has the further advantage that large plasmids are

1

readily visualized. It has been used successfully with a variety of bacterial genera, including enteric species as well as species of *Neisseria*, *Staphylococcus*, *Pseudomonas*, *Rhizobium*, *Agrobacterium*, *Clostridium*, and *Campylobacter*.

A. SMALL CULTURE VOLUMES (1–10 mL)

PROCEDURE

1. Grow cells in enriched medium. Aerated cultures should be grown to late exponential phase while nonaerated cultures can be incubated overnight (O/N).

2. Pellet the cells by centrifugation for 10 minutes, at 25°C, at full speed in a clinical centrifuge. Remove the supernatant and resuspend the cells in 1 mL of TE buffer. Transfer the cells to a 1.5 mL microfuge tube and pellet by centrifugation for 2 minutes at 4°C at full speed in a microfuge. After removing the supernatant, resuspend the cells in 25 μL of TE buffer and vortex vigorously.

3. Add 400 μL of lysis buffer (TE buffer at pH 12.45 containing 1.5% sodium dodecyl sulfate, SDS).

COMMENTS

1. For enteric species, grow the cells at 37°C in 1–3 mL of trypticase soy broth (any enriched media can be used). With *Clostridium* species, cells should be grown anaerobically in 5 mL of brain heart infusion broth for 36–48 hours. Subinhibitory concentrations of antibiotic (as determined by the resistances specified by the plasmid) may be added to the medium as a means of inhibiting the loss of unstable plasmids. With *Neisseria* species (and with the enterics), a quadrant of growth from solid medium that has been supplemented with antibiotic may be directly suspended in 500 μL of TE buffer.

TE Buffer

50 mM Tris-HCl, pH 8.0
20 mM ethylenediaminetetraacetic acid (EDTA)

2. The resuspension of cells in TE buffer gives much better lysis than the direct treatment of pelleted cells.

3. In a recent modification Birnboim (1983) has suggested using lysozyme treatment before the alkaline-SDS step; however, most cells lyse well without this.

LYSIS BUFFER: The alkaline-SDS solution should be freshly prepared every 2–3 days and should be stored in a screw-cap bottle at room temperature (at 4°C, SDS comes out of solution). The solution absorbs CO_2 from the air upon standing, causing the pH to decrease.

Dissolve the SDS in TE buffer, pH 8.0 (SDS and other lysis chemicals are reagent grade). Adjust the pH to 12.0 with 10N NaOH and to pH 12.45 with 3N NaOH. The inclusion of 50 mM glucose in the SDS solution is reported to facilitate pH adjustment Birnboim, 1983).
At a pH of 12.0 to 12.5, linear and chromosomal DNA are selectively denatured while covalently closed

circular (CCC) DNA is not affected. However, at greater pH values CCC-DNA may become irreversibly denatured. Proteins are also denatured at pH 12.5 thereby reducing the possibility of enzymatic degradation of plasmid DNA. The SDS lyses the cells and denatures proteins.

4. Mix the suspension by inverting the microfuge tube at least six times. Let the lysis mixture stand either for 30 minutes at room temperature or for 60 minutes at 56°C. Mix the tubes by inversion at least once during the incubation period.

4. Adequate lysis is achieved after 30 minutes at room temperature; however, incubation at 56°C enhances the denaturation of chromosomal DNA thereby giving "cleaner" preparations.

5. Neutralize the mixture by adding 150 μL of 3M sodium acetate, pH 4.8. Gently mix several times by inversion.

5. Adjust the pH of the 3M sodium acetate with acetic acid.

Lowering the pH causes chromosomal DNA to form an insoluble complex that, together with protein-SDS complexes, is precipitated by the high salt concentration. Substitution of potassium acetate for sodium acetate may enhance precipitation (Birnboim, 1983). In this case the pH is adjusted with formic acid.

With genera containing high levels of nuclease, the nuclease inhibitor diethyl pyrocarbonate may be added in a ratio of 1:500 (inhibitor: lysate) immediately following neutralization.

6. Place the suspension on ice for 60 minutes to complete precipitation of high molecular weight complexes.

7. Centrifuge for 15 minutes at full speed in a microfuge. Recover the supernatant using a polyethylene Pasteur pipette and place in a clean microfuge tube.

7. Frequently, a rather "fluffy" pellet is obtained. In this instance, increased yields of DNA may be achieved by recentrifuging the tube after removing the supernatant to a second tube. Then remove the supernatant and combine with that from the initial centrifugation.

Note: Use polyethylene pipettes, since DNA may stick to the walls of glass pipettes.

8. Phenol/Ether Extractions

(Optional)

8. The decision to include these steps is determined by the ultimate use of the DNA. They are not required for either plasmid screening or transformation procedures.

PHENOL EXTRACTION: Phenol denatures proteins and selectively removes denatured DNA from aqueous solutions. It is therefore frequently used to "clean-up" cell lysates. The phenol used is saturated with a solution of the major components of the aqueous phase (e.g., salts, buffer, H$_2$O), so as to prevent their excessive loss from solution during the extraction (see Section XVII-F of this chapter).

a. Add 0.5 mL of sodium acetate (1M) saturated phenol to the neutralized mixture (see Section XVII-F for the preparation of phenol solutions).

b. Invert the tube several times in order to mix the two phases.

c. Centrifuge the tubes in a microfuge for 10 minutes, at full speed, at 4°C.

d. Remove the UPPER aqueous phase to a clean microfuge tube using a polyethylene Pasteur pipette.

PROCEDURE

e. Add an equal volume of reagent-grade ether to the aqueous phase. Mix by inversion.

f. Allow the phases to separate; then discard the upper ether layer. The DNA remains in the LOWER layer.

g. Repeat 2–3 times.

h. Remove residual ether by evaporation under an air stream.

9. Precipitate the DNA by adding 2 volumes of cold ethanol to the supernatant. Mix by inversion and store at −20°C for 2 hours.

10. Pellet the DNA by centrifuging the tubes at full speed in a microfuge for 10 minutes at 4°C.

11. Pour off the ethanol, invert the tube, and air dry.

12. Add 20 to 100 μL of the appropriate buffer. Make sure that the pellet is completely dissolved.

COMMENTS

ETHER EXTRACTION: Because trace amounts of phenol may interfere with the activity of certain restriction endonucleases and other enzymes, ether is used to remove residual phenol from the aqueous phase.

9. DNA may also be precipitated O/N at −20°C, or for 20 minutes at −70°C; or it may be placed in a dry ice–ethanol bath for about 5 minutes. The 2 volumes of ethanol may be replaced with 1 volume of isopropanol.

10. The pellet may or may not be visible after centrifuging. If using a fixed-angle rotor, it is advisable to place the tubes in a specific orientation in the centrifuge (e.g., cap attachment upwards) so that the position of the (invisible) pellet is known during the resuspension step.

12. The buffer employed depends on the subsequent use of the DNA. For example, for restriction endonuclease analysis, restriction buffer (Section IX) will be used, while TE buffer may be used for screening DNA by agarose gel electrophoresis. The method outlined above produces enough DNA for several restriction endonuclease anlayses.

B. MODERATE CULTURE VOLUMES (50 mL)

PROCEDURE

1. Grow the cells O/N in 30 to 50 mL of trypticase soy broth under the appropriate conditions.

2. Pellet the cells by centrifugation (12,000 × *g*, 15 minutes, 4°C). Wash with 10 mL of TE buffer.

3. Pellet the cells as described in step 2. Resuspend the pellet in 0.5 mL TE buffer and place the suspension in a 50 mL polypropylene centrifuge tube.

COMMENTS

1. Provided that the ratio between reagents is maintained, this procedure may also be scaled down to small (microfuge) volumes. Extracts prepared using this protocol display less background RNA and sharper plasmid bands in agarose than those prepared by Method 1A (Bezanson *et al.*, 1982).

3. Mix to ensure that the suspension is completely homogeneous. Polypropylene or glass centrifuge tubes are used because these materials are resistant to phenol.

Alternatively, steps 3 to 8 can be completed in 50 mL glass beakers, and mixing accomplished using magnetic stirring bars.

4. Add 9.5 mL of lysis buffer (TE buffer, 1.5% SDS, pH 12.45). Mix by gently swirling the contents of the tube.

4. Prepare lysis buffer as outlined in Method 1A, p. 2.

5. Incubate for 1 hour at 56°C or for 30 minutes at room temperature.

5. The elevated temperature coupled with the alkaline pH effects denaturation of chromosomal DNA without affecting plasmid DNA.

6. Add 0.8 mL of 2M Tris-HCl (pH 7.4). Swirl gently to mix. Test with pH paper.

6. It is essential to lower the pH of the suspension, or else the removal of the upper aqueous layer after phenol extraction (see step 8) is difficult (the layer is "sticky" and phenol may also be simultaneously removed). After adding 2M Tris-HCl buffer the pH should drop to about 8.75, although the volume of buffer required for neutralization varies with different bacterial genera.

7. Add 1.0 mL of 5N NaCl (final concentration, 3%). Mix with a gentle swirling motion. Let stand at 4°C for 1 hour.

7. The high salt concentration is used to precipitate high molecular weight nucleic acid and protein-SDS complexes.

8. Add 10 mL of phenol saturated with 3% NaCl (see Section XVII-F of this chapter). Mix by gentle inversion.

8. Phenol denatures and extracts proteins. The mixture will separate into an upper aqueous phase, and the lower, hydrophobic, phenol phase.

9. Centrifuge for 10 minutes at 5000 × g (4°C). Remove the upper phase with a polyethylene pipette and place in a clean polypropylene tube. Note the volume.

9. Generally, about 8–9 mL of the aqueous phase is recovered.

10. Adjust the supernatant to 0.3M sodium acetate by adding an appropriate volume of 3M stock solution.

10. The addition of the monovalent cations, Na^+, NH_4^+, or K^+, as acetate or chloride salts (to final concentrations of 0.3M, 3.7M, and 0.1M, respectively) selectively alters the solubility of DNA, thereby increasing the efficiency of subsequent alcohol precipitations. The calculation for determining the volume of salt (e.g., sodium acetate) to add is as follows

$$x = \frac{0.3(y+x)}{3}$$

Where x = volume of stock solution (3M sodium acetate) to be added

y = existing volume

Thus, if 10 mL is recovered from the aqueous phase, 1.1 mL of 3M sodium acetate (1 mL will do) should be added to adjust the concentration to 0.3M.

Use stock solutions of 1M potassium acetate and 7.5M ammonium acetate. Note that the ammonium acetate

PROCEDURE

11. Add 2 volumes of cold ethanol or 1 volume of cold isopropanol to the solution to precipitate the DNA. Mix by gentle inversion, then store at −20°C O/N.

12. Pellet the DNA by centrifugation for 30 minutes at −10°C and 35,000 × g. Pour off the supernatant and invert tubes to air dry the pellet.

13. Dissolve the DNA in 100 μL of the appropriate buffer.

COMMENTS

procedure is of limited utility in microfuge-scale preparations because of the large volumes needed to achieve the desired final concentration of 3.7M.

12. Use a microfuge for smaller volumes (full speed, 10−12 minutes, 4°C). The pellet may not be visible if a microfuge procedure was followed.

METHOD 2: BOILING PROCEDURE

This procedure is an adaptation of the method developed by Holmes and Quigley (1981). It is ideal for screening large numbers of colonies for plasmid content; plasmids ranging in size from 3.2 to 39.2 kilobases (kb) are easily visualized. In the hands of the novice, however, this method may cause many frustrations and equivocal results. Therefore, we recommend that beginners start with a method such as the alkaline detergent procedure to gain confidence in their technical skills. The major difficulty in using the boiling method is in ascertaining whether the cell suspension is adequate (too few or too many cells will give equivocal results). Since the density of cells required for lysis varies with different genera (we have routinely used this method for identifying plasmids in *Neisseria* species and *Escherichia coli*, but have had less success in lysing anaerobes and enteric isolates harboring large plasmids). Only practice will ensure reproducible results.

PROCEDURE

1. Depending on the genus to be lysed (some genera grow less well than others), suspend either a single colony (4 mm), or a loopful of cells from solid medium in 500 μL of STET buffer in a 1.5 mL microfuge tube. Vortex to ensure an even, homogeneous suspension.

COMMENTS

1. One colony of *E. coli* will give a reasonable suspension; on the other hand a loopful of *N. gonorrhoeae* cells is required to achieve the same cell density. Alternatively, cells may be pelleted from a nonaerated O/N broth culture (3−5 mL) and then resuspended in 0.5 mL STET buffer.

STET Buffer

8% sucrose
5% Triton X-100
50 mM EDTA
50 mM Tris-HCl, pH 8.0

2. Add 100 μL of freshly prepared lysozyme (10 mg/mL). Vortex mix. Let stand for 10 minutes at room temperature. Lysis can be monitored visually (increase in viscosity, clearing).

2. Occasionally, good lysis is achieved without this step; however, cultures should not be more than 18 hours old. The lysozyme step is a tricky one. With very high concentrations of lysozyme, cells may not lyse. In

PROCEDURE	COMMENTS
	addition, the enzyme is pH-dependent and may not work well if the pH is less than 8.0.
3. Place the suspension in a boiling water bath for 60 seconds.	**3.** Boiling temperatures denature nucleases that may be present in the extract. Longer periods of boiling may result in the appearance of chromosomal DNA in final preparations due to the fragmentation of the long strands.
4. IMMEDIATELY centrifuge the suspension in a microfuge at maximum speed (e.g., Eppendorf, 15,600 × g) for 15 minutes.	**4.** After centrifugation, a fluffy (not densely packed), whitish pellet generally indicates that good lysis has been achieved. The pellet comprises chromosomal DNA and denatured proteins.
5. Transfer the supernatant to a clean 1.5 mL microfuge tube.	**5.** If purifying DNA from *Neisseria* species, a **PHENOL EXTRACTION** step should be added at this point (Dillon *et al.*, 1983). TE-saturated phenol should be used.
6. Precipitate the DNA by adding either 2 volumes of cold ethanol or an equal volume of cold isopropanol. Store at −20°C for 2 hours or at −70°C for 10−20 minutes.	**6.** DNA may be precipitated O/N at −20°C if desired. The solution may be cloudy or slightly frozen, but this will not affect the final result.
7. Pellet the precipitated DNA by centrifuging for 15 minutes as described in Step 4.	**7.** The DNA pellet may not be visible.
8. Decant the alcohol. Air dry the pellet for about 10 minutes or until all of the alcohol has evaporated.	
9. Resuspend the pellet in 50−100 μL of the appropriate buffer.	**9.** DNA prepared using this method may be used for transformation and may be screened by agarose gel electrophoresis. There is generally enough DNA for several restriction enzyme digests.

II. Isolation of Plasmid DNA by Cesium Chloride–Ethidium Bromide (CsCl–EtBr) Ultracentrifugation or Column Chromatography

A. ISOLATION OF PLASMID DNA FROM LARGE CULTURE VOLUMES (> 200 mL) ON CESIUM CHLORIDE GRADIENTS

A number of procedures (e.g., digestion with restriction endonucleases, nick translation, DNA sequencing) require higher concentrations of plasmid DNA, free of contaminating molecules (chromosomal DNA, high molecular weight RNA, protein). A variety of methods have been developed for obtaining pure plasmid DNA, including scaled-up versions of the boiling method and the alkaline detergent method (Holmes and Quigley, 1981; Casse *et al.*, 1979). In the method described below, the cells are lysed with the nonionic detergent Triton-X100, and the resultant cleared lysate is centrifuged through a cesium chloride–ethidium bromide gradient. This method is a modification of the method developed by Clewell and Helinski (1969) and later amended by Goebel (1970).

PROCEDURE	COMMENTS
1. Grow cells O/N in 20 mL of the appropriate growth medium.	**1.** In cases where certain bacteria do not grow well in liquid culture (e.g., *N. gonorrhoeae*), suspend

growth from three to five agar plates (100×15 mm) in TES-sucrose and begin at step 4. If R-plasmids are being isolated, subinhibitory concentrations of antibiotic (as determined by the R-type) may be added to the growth medium to assure a high percentage of R+ cells.

2. Dilute the cells 1:20 into 200 mL of fresh broth in a 1 L flask. Incubate, with aeration, at 37°C for 45 to 60 minutes (so as to achieve midlog phase).

2. Provided that plasmid replication does not require new protein synthesis (i.e., not inhibited by chloramphenicol), the **AMPLIFICATION** of plasmid DNA may be achieved through the addition, at midlog phase, of chloramphenicol powder (150 µg/mL), followed by a further 3 to 5 hour incubation period. Examples of two plasmids that may be amplified are pBR322 and *Col*E1

3. Harvest the cells by centrifugation at 4°C for 15 minutes at $10,000 \times g$.

3. The cells may be washed once with TES buffer, if desired.

4. Resuspend the pellet in 1.5 mL of TES (pH 8.0) buffer containing 25% sucrose (w/v). Vortex mix. Add sufficient TES-sucrose to achieve a final volume of 2.0 mL.

4. TES Buffer

0.03M Tris-HCl, pH 8.0
0.05M EDTA
0.005M NaCl

Experience will allow one to determine the ideal concentration of cells. If too many cells are lysed, difficulty will be experienced in pelleting cell debris during the clearing spin.

5. Add 0.25 mL of lysozyme (10 mg/mL in TES, pH 8.0). Mix by inversion and place the solution on ice for 15 minutes.

6. Add 0.2 mL EDTA (0.25M) and mix by inversion.

6. EDTA, a chelating agent, removes from solutions metal ions that are essential for the activity of certain nucleases. It thus protects DNA from digestion with nucleases.

7. Add 0.7 mL of 20% Triton-X100 solution. Mix by inversion and place on ice for 10 minutes. After this period check for a change in viscosity.

7. Other detergents may be used to produce a cleared lysate. However, if DNA is to be purified on CsCl gradients, one should not use sodium dodecyl sulphate, as it is insoluble in CsCl. Detergents which may be used include sodium sarcosinate and Brij 58.

8. Centrifuge the suspension at $37,000 \times g$ for 20 minutes, at 4°C.

9. Remove the supernatant with a polypropylene pipette and place in a *plastic* tube or beaker. Note the volume collected.

10. Adjust to a final volume of 4.0 mL with TES buffer. Add 3.65 g of cesium chloride. Completely dissolve the cesium chloride by gently swirling the mixture.

10. One may also add a small volume of EDTA (0.5 mL, 0.25M) to inhibit nucleases that may be present in CsCl.

11. Add 50–100 µL of **ETHIDIUM BROMIDE SOLUTION** (10 mg/mL in H₂O) and gently mix.

11. *CAUTION: Ethidium bromide is a potent mutagen. Disposable vinyl gloves should be worn at all times when this agent is handled.*

PROCEDURE

COMMENTS

Since visible light may cause single-stranded nicks in the DNA in the presence of ethidium bromide (Clayton *et al.*, 1970), darken the room or wrap each container in foil before addition of the dye.

Ethidium bromide intercalates with DNA, thereby decreasing its density in CsCl. Covalently closed circular (CCC) DNA binds much less with ethidium bromide than open circular (OC) and linear DNA. Therefore, CCC-DNA can readily be separated from other types of DNA on the gradient (Radloff *et al.*, 1967).

12. Pour the DNA solution into an ultracentrifuge tube and add light mineral oil to the top of the tube.

12. The oil is added to prevent the centrifuge tubes from collapsing.

13. Balance the tubes with mineral oil, then cover or seal as specified by their manufacturer. Place in the appropriate rotor and spin.

14. Conditions for centrifugation in a Beckman L5-65 ultracentrifuge are given in Table 1.1.

TABLE 1.1 Conditions for Running CsCl Gradients in a Beckman L5-65 Ultracentrifuge

Rotor Type	Speed in RPM	Time Required	Temperature
Type 50	44,000	40–44 hours	5°C
Type 65	44,000	40 hours	5°C
VTi 65	50,000	20–24 hours	5°C

15. Recover the plasmid band as indicated in Section III, p. 11.

16. Remove ethidium bromide from the DNA solution as described in Section IV, p. 12.

17. Adjust the solution to 0.3M sodium acetate; then precipitate the DNA with ethanol.

17. Use procedures outlined in steps 10 and 11 (Method 1B), Section I, p. 5.

18. Resuspend the DNA in 100 μL of the appropriate buffer.

18. DNA concentration may be determined as described in Section XVII-M, p. 102.

B. SEPARATION OF PLASMID DNA USING AFFINITY OR ION-EXCHANGE CHROMATOGRAPHY

Elution of particulate material through a vertical column provides an alternative to electrophoresis and cesium chloride centrifugation for the isolation, separation, and/or purification of DNA. As with gel electrophoresis (Section V, p. 13), column chromatography is rapid, gentle, displays high resolving power, and gives reproducible results. Unlike gel systems however, it allows for the quantitative recovery of DNA (> 90%). Column chromatography yields DNA preparations of relatively high purity, (e.g., suitable substrates for restriction endonuclease digestion), as does cesium chloride–ethidium bromide gradients, but is less time consuming, less expensive, has a greater sample capacity, and displays a superior resolving power. Unlike both gel electrophoresis and cesium chloride centrifugation, chromatography materials are reusable. Affinity adsorbents and ion-exchange resins are chromatography materials of particular utility in nucleic acid work.

Affinity chromatography is a form of adsorption chromatography in which materials are separated by utilizing biospecific interactions, such as antigen-antibody reactions or the intercalation of nucleic acids with dyes. Affinity-based separation of nucleic acids may be achieved with acridine yellow or phenol neutral red-conjugated bisacrylamide (Bünemann and Müller, 1978). The differential intercalation of the immobilized dyes into the passing nucleic acids allows the separation of supercoiled DNA from open circular and linear forms. The DNA is eluted, using a gradient of increasing sodium perchlorate concentration (NaCl may also be used but with a decrease in peak sharpness). The manufacturer (Boehringer-Mannheim) claims a capacity for such columns of 40 μg DNA per mL of settled gel. Molecules up to 100 Md in mass have been purified in such columns. These columns have also been used to isolate plasmid DNA from the other cellular components present in detergent lysates (Vincent and Goldstein, 1981).

Dye-based affinity gels have the disadvantage of being light-sensitive, necessitating the use of protective films when they are being run coupled with storage in the dark. Further, the DNA solution placed on the column must be relatively pure. Proteins, and to some extent RNA, will interact with the dyes, thereby decreasing their affinity for DNA. Protein may be removed by phenol-extraction or hydroxylapatite chromatography (see subsequent paragraphs). Of minor consideration is the need for relatively high column pressures (i.e., peristaltic pumps) to achieve reasonable flow rates.

Ion-exchange chromatography of nucleic acids is based on the interaction between the strongly acidic phosphate residues of the polynucleotides and anion exchangers immobilized in the column matrix. During this interaction nucleic acid becomes tightly bound to the chromatography resin. Once bound, the nucleic acid can be removed only by increasing the ionic strength of the chromatography slurry to a value higher than that required for binding (gradient elution). This is accomplished by washing the column with a gradient of increasing salt (NaCl, KCl) concentration.

One such nucleic acid chromatography system uses trialkylmethylammonium chloride as an anion exchanger. A thin film of this compound is formed over solid, nonporous, and chemically inert particles of various sizes, to give the chromatography resin. This system binds nucleic acids at low salt concentrations (0.1−0.5N NaCl) and releases them at high concentrations (0.7−2.0N NaCl). Single-stranded DNA is bound more tightly than duplex DNA, larger polynucleotides more than smaller polynucleotides, and linear molecules more than circular ones of the same size. The general purpose resin NACS-52, is reported by its manufacturer (Bethesda Research Laboratories) to have a binding capacity of 0.25 mg DNA per gram of resin. This system has been used with success to isolate plasmid DNA from cell lysates (Best *et al.*, 1981) and to fractionate restriction endonuclease-generated DNA fragments (Hardies *et al.*, 1979). Other uses, such as the removal of unreacted molecules from *in vitro* systems and the extraction of nucleic acids from electrophoresis gels, have been suggested for this system (Thompson *et al.*, 1983). As with affinity absorbents, a number of the nucleic acid chromatography resins require the use of pressure-flow devices. One resin, NACS-52, because of its particle size requires gravity flow only. As with the affinity systems, DNA preparations must be relatively clean and free of phenol, ethanol, and detergents prior to being loaded onto the column.

Hydroxylapatite is a popular ion-exchange resin. Columns formed with this calcium phosphate derivative facilitate the rapid isolation of plasmid DNA (Colman *et al.*, 1978), the separation of single- and double-stranded DNA (Miyazawa and Thomas, 1965), the deproteinization of DNA solutions (Vincent and Goldstein, 1981), and the recovery of DNA from electrophoresis gels (Hansen, 1976; Tabak and

Flavell, 1978). Nucleic acids are bound to the column as a result of the interaction of their phosphate residues and the calcium of the adsorbent. They are then recovered by elution with a phosphate gradient. The capacity of hydroxylapatite is approximately 1.5 mg DNA per gram of resin at pH 6.8. Single-stranded DNA displays a lower affinity for hydroxylapatite than does double-stranded DNA, due largely to reduced amounts of phosphate residues and configurational differences. Reasonable flow rates are achieved by the use of pressure.

Since all of the chromatography material described above may be reused (regenerated) and since their columns give reproducible separations, the gradual alteration in ionic conditions achieved by gradient elutions can be replaced in most instances by more radical, stepwise changes in ion concentration (step elutions). Hence, one can proceed directly to the conditions ideal for the elution of a particular molecule, thereby greatly speeding up separations and permitting the use of so-called mini-columns of very short lengths. (For details on column formation and use, see Section XVII-N, p. 103).

III. Recovery of Plasmid DNA from CsCl−EtBr Solutions

PROCEDURE

COMMENTS

CAUTION: Protective eyewear should be worn when visualizing bands under uv.

1. To locate the plasmid band, place the centrifuge tube under a longwave ultraviolet light source. Normally two bands are present: closed-circular plasmid DNA forms the lower band; open-circular plasmid and chromosomal DNA form the upper. Mark the positions of the bands with a felt pen.

1. Ethidium-bromide-DNA complexes fluoresce if illuminated by uv light sources of 254, 300 or 366 nm (Brunk and Simpson, 1977; Southern, 1979). Because the shorter wavelengths may damage the DNA by nicking or dimerization, a longwave (366 nm) uv light source is used. Bands appear as reddish lines in an orange-pink background. Under some conditions bands may be detected by the naked eye.

With some procedures designed to eliminate chromosomal DNA, only one band may be seen.

2. Remove the lower band as follows:

a. Remove the cap, or in the case of vertical rotor tubes, prick a small hole at the top of the tube to open the system to atmospheric pressure.

b. Puncture the tube at the side, just under the plasmid band, with a 16 or 21 gauge needle (bevel up). The band may be removed with a syringe or by collecting drops through the end of the needle. Alternatively, drops can be collected by puncturing the bottom of the tube.

2. The pellet at the bottom of the tube consists of high molecular weight RNA. Low molecular weight RNAs do not pellet and are dispersed throughout the tube. Protein material is often visualized as a pellicle at the oil/CsCl interface.

If side puncture is used, be careful not to punch a hole through to the other side of the tube. Place a beaker under the tube to collect overflow.

Be *sure* not to contaminate the plasmid band with the upper band (if the boiling method was used to prepare plasmid DNA, the upper band may not be present). To ensure a pure preparation, it may be necessary to rerun

PROCEDURE	COMMENTS
	the plasmid band on another CsCl gradient. In any event, check the band for purity by running the DNA on an agarose gel (Section V, p. 13).

IV. Removal of Ethidium Bromide from DNA Solutions

METHOD 1: ALCOHOL EXTRACTION

PROCEDURE	COMMENTS
1. Add 1 volume of buffer-saturated isopropanol or isobutanol to the DNA solution.	**1.** Both isopropanol and isobutanol can concentrate DNA by removing water from the suspension. Therefore, to avoid changes in the volume of the buffer system, it is necessary to use isopropanol or isobutanol saturated with the buffer in which the DNA is suspended.
2. Mix gently by inversion. Allow the phases to separate and remove the upper phase.	**2.** Ethidium bromide is extracted into the upper organic phase, which appears pinkish.
3. Repeat at least one time.	**3.** It is essential to remove all ethidium bromide as residual amounts may interfere with restriction endonuclease digests (Parker *et al.*, 1977; Osterlund *et al.*, 1982).

METHOD 2: EXTRACTION ON A DOWEX COLUMN

1. Place a wad of siliconized glass wool at the junction of the wide and narrow portions of a (siliconized) Pasteur pipette. Attach a short length of small-gauge plastic tubing to the tip of the pipette. Clamp the assembly in a vertical position so that the tubing is at the bottom (column exit).	**1.** See Section XVII-H (p. 100) for siliconization procedure.
2. Pour a water slurry of Dowex 50 × 8 (50−100 mesh) into the Pasteur pipette to form a column about 4 cm in length. Neutralize by washing extensively with distilled H_2O.	**2.** Dowex 50 is a strongly acidic, polystyrene cation-exchange resin. Nucleic acids having the same charge as the resin's functional group are not adsorbed onto the resin but remain in solution.
	Washing is accomplished by repeated additions of elutant to the top (wide portion) of the pipette. A clamp on the plastic tubing will control flow rates.

This method is modified from that of Radloff *et al.* (1967) and is sometimes used with radioactively labeled DNA.

An **alternative method with Dowex** (W. Goebel, personal communication to G. Bezanson) is the following.

a. Pour the Dowex 50 × 8 slurry into a 1 mL plastic syringe which has been plugged with a small wad of siliconized glass wool.

b. Load the DNA sample and then place the syringe in an 8−10 mL centrifuge tube.

c. Spin at 10,000 × *g* for 10 minutes at 4°C.

d. Remove the DNA solution from the bottom of the centrifuge tube.

3. Load the DNA sample.

4. Wash with TES buffer (pH 8.0) and collect 0.2 to 0.5 mL fractions.

4. The column can be used several times. After each use, rinse extensively with water.

5. If the DNA is radioactively labeled, spot 10 μL of each fraction onto filter paper and fix with 10% cold trichloroacetic acid.

5. See Section XVII-O, p. 106, for information regarding TCA-fixation and liquid scintillation counting.

If the DNA is not radioactively labeled, measure the optical density at 260 nm.

6. Determine the radioactive peak in a scintillation counter.

6. See Section XVII-O, p.106.

V. Gel Electrophoresis

One of the most powerful and commonly used techniques for the separation of nucleic acids according to their mass and structural configuration is electrophoresis in either agarose or acrylamide gels (Loening, 1967; Aaij and Borst, 1972; Meyers *et al.*, 1976; Sealey and Southern, 1982), or a combination of the two (Peacock and Dingman, 1968). Duplex molecules, such as covalently closed-circular plasmid DNAs, move more rapidly through the gel matrix toward the anode than do larger, less compact molecules such as open-circular or chromosomal DNAs. The resolving power of the gel depends on the concentration of material used, which in turn determines the pore size of the gel. For example, a 2% agarose gel can resolve duplex molecules as small as 500 base pairs. For large DNA fragments a low concentration agarose gel (0.3% to 0.1%) is used. Polyacrylamide gels are used for separating fragments between 6 (20% polyacrylamide) and 1000 (3% polyacrylamide) base pairs in length, as well as for separating single-stranded DNA. However, when separating single-stranded DNA or RNA, it is preferable to use denaturing electrophoresis conditions. Denaturing gels and buffers generally contain urea, alkali, or other denaturants (see Section XV, p. 88).

A. APPARATUS

An inexpensive apparatus that is easily "homemade" and that accommodates the electrophoresis of both horizontal and vertical gels, as well as mini-gels, is illustrated in Figures 1.1 and 1.2. The basic apparatus for horizontal gels consists of the electrophoresis chamber with its central supporting block (Figure 1.1*a* and 1.1*b*), together with the back plate and associated spacers (dimensions given in Figure 1.1*c*). All are constructed of clear acrylic plastic (Plexiglas). Ultraviolet-transparent (uvt) Plexiglas may be used so that gels can be viewed or photographed on a transilluminator without removal from the plate. The central supporting block may be cooled by passing water through the portals on either side (for most purposes cooling is not necessary). The apparatus is assembled using a glue made of leftover Plexiglas dissolved in methanol. (Acrylic plastic takes two to three weeks to dissolve in the methanol). The platinum wire used in the apparatus is 30 gauge and 0.25 mm in diameter.

The vertical gel electrophoresis apparatus utilizes the electrophoresis chamber from the horizontal apparatus to form the lower buffer chamber. The upper buffer chamber (Figure 1.2*a*) is placed on the central supporting block (see Figure 1.2*b*). Vertical gels may be cooled by running water through the upper chamber portion of the apparatus. The dimensions of the plates and spacers used for vertical agarose gels are indicated in Figure 1.2*c*.

The well-forming comb (Figure 1.3) may be used in preparing both horizontal and vertical gels. Care should be taken to ensure that the comb does not have rough edges which may cause the gels to tear.

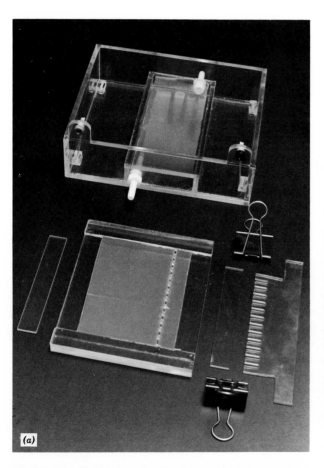 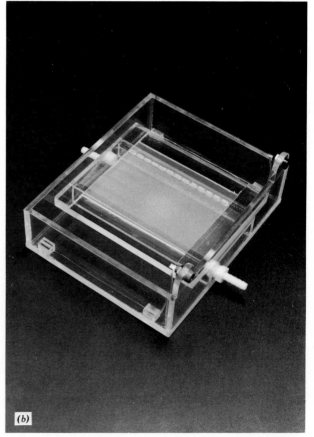

FIGURE 1.1 (*a* and *b*) Horizontal gel electrophoresis apparatus.

ELECTROPHORESIS CHAMBER

PLATES FOR CASTING GELS

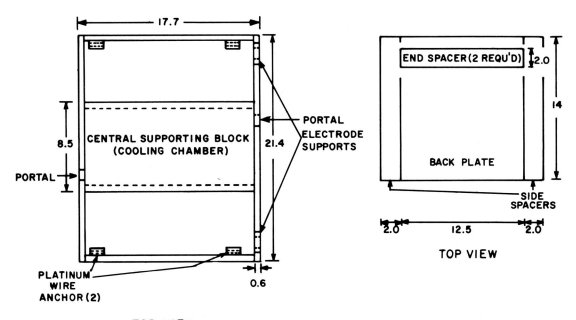

TOP VIEW

TOP VIEW

SIDE VIEW

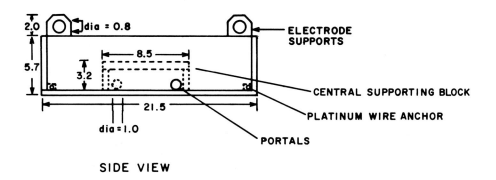

SIDE VIEW

1) ALL DIMENSIONS IN CM
2) SEE FIG. 1A OR 1B FOR POSITION OF PLATINUM WIRE

FIGURE 1.1c Apparatus Dimensions For Horizontal Gels.

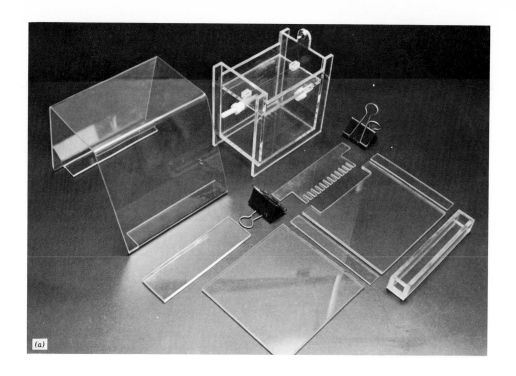

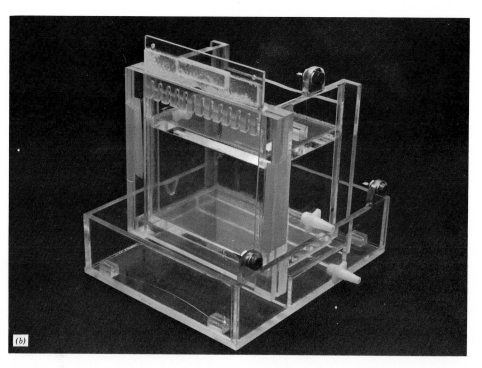

FIGURE 1.2 (*a* and *b*)   Vertical gel electrophoresis apparatus.

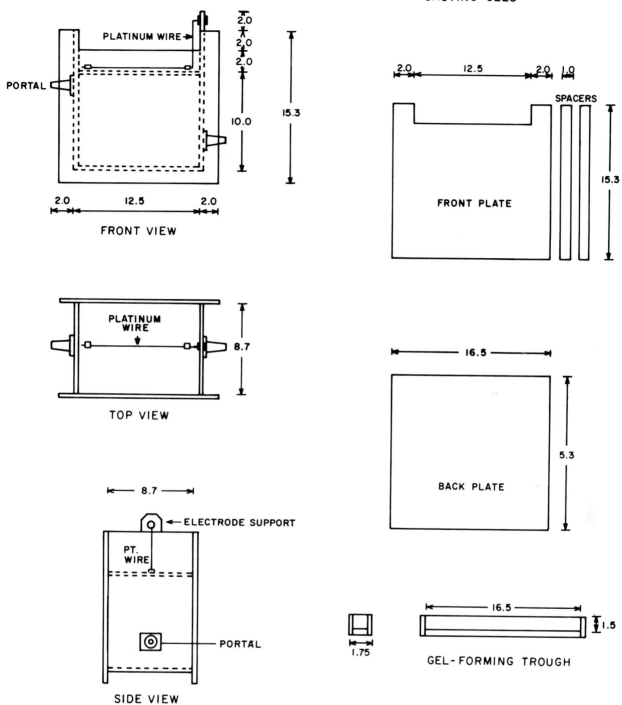

UPPER BUFFER CHAMBER

PLATINUM WIRE

PORTAL

2.0
2.0
2.0
10.0
15.3

2.0 12.5 2.0

FRONT VIEW

PLATINUM WIRE

8.7

TOP VIEW

8.7

ELECTRODE SUPPORT

PT. WIRE

PORTAL

SIDE VIEW

PLATES AND SPACERS FOR CASTING GELS

2.0 12.5 2.0 1.0

SPACERS

15.3

FRONT PLATE

16.5

5.3

BACK PLATE

16.5 1.5

1.75

GEL-FORMING TROUGH

1) ALL DIMENSIONS IN CM
2) SPACERS 0.3 CM THICK FOR AGAROSE GELS, 0.15 CM THICK FOR POLYACRYLAMIDE GELS

FIGURE 1.2c Apparatus dimensions for vertical gels.

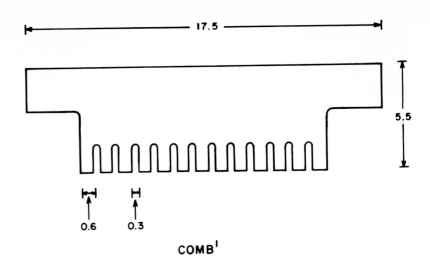

COMB[1]

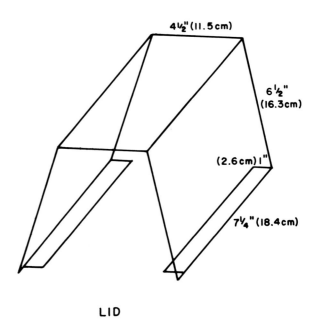

LID

1) THICKNESS OF COMB = 0.3 CM FOR AGAROSE GEL
0.15 CM FOR POLYACRYLAMIDE GEL

FIGURE 1.3   Comb and lid for horizontal and vertical gel electrophoresis apparatus.

METHOD 1:   HORIZONTAL AGAROSE GELS

Agarose gels can be run either horizontally or vertically. Horizontal agarose gels are generally used for "sloppy" (under 0.8% agarose) gels and can be run under the buffer ("submarine gels"), thereby providing phase continuity and diminishing the effects of temperature gradients and electrical field discontinuities.

PROCEDURE	COMMENTS

1. As indicated in Figure 1.1a, one base plate with two side spacers, and two end spacers are required to prepare horizontal gels. The side spacers are glued to the base plate.

1. The thickness of the gel will depend on the thickness of the spacers and the volume of agarose poured. Generally, agarose gels less than 3 mm thick are too fragile to handle.

2. Wash the plate and the end spacers with ethanol to remove any grease and debris from previous gels.

3. Position the end spacers between the side spacers. Use masking tape to attach the end spacers to the side spacers.

3. The length of the gel can be varied by adjusting the position of the end spacers on the base plate. Full-length gels may be prepared by replacing the end spacers with water resistant tape at the edges of the plate.

If a spacer is glued or taped in the middle of the base plate, two **MINI-GELS** of different agarose concentration can be run at the same time (Figure 1.4). The length of the mini-gel can be adjusted by varying the position of the end spacers. The volume of agarose needed for each mini-gel is about 10–20 mL, depending on the length of the gels.

4. Prior to casting the gel, the height of the well-forming comb must be adjusted so that the bottom of the comb does not touch the base plate. This is done by positioning the comb with two clamps which rest on the the side spacers. Set the comb aside, before pouring the gel.

4. If the comb touches the base plate an imperfect well will be formed and the DNA sample will leak underneath the gel.

5. Prepare a solution of molten agarose (50 mL when using apparatus described in Figure 1.1a) by adding the appropriate amount of agarose powder to 1× electrophoresis buffer. Bring the solution to a boil. The agarose must be completely dissolved prior to pouring.

5. The agarose solution can be rapidly brought to a boil in a microwave oven or over a hot plate.

Several types of **agarose** are available commercially. For general screening, standard low-endoosmotic agarose is adequate and economical. However, in procedures that require elution of DNA from the gel

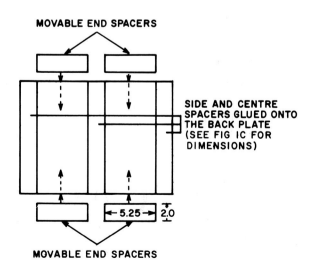

FIGURE 1.4 Apparatus for mini-gels.

and/or the use of certain enzymes (e.g., ligases, restriction endonucleases) it is recommended that an ultra-pure agarose be used. This grade of agarose is free of agaropectin, a sulfated oligosaccharide that is a common contaminant of agarose that binds to nucleic acids and that can inhibit enzymes. In addition, high purity agarose generally exhibits minimal uv quenching, thereby increasing the sensitiviity of DNA visualization after staining with ethidium bromide.

ELECTROPHORESIS BUFFERS: the two most commonly used buffers for agarose gel electrophoresis are tris-acetate (Loening, 1967) and tris-borate (Peacock and Dingman, 1968). The buffers are generally stored as 10× solutions and are diluted to 1×, as required. The mobility of linear and open-circular (OC) DNA in agarose gels is different when the two buffers are used. In tris-acetate buffer, linear DNA runs before OC-DNA, while with tris-borate buffer the running order is reversed.

10× Tris-acetate Buffer

400 mM Tris base
200 mM sodium acetate
 18 mM EDTA
pH 7.8

10× Tris-borate Buffer

890 mM Tris base
890 mM boric acid
 25 mM EDTA
pH 8.3

For general screening purposes, a 1% agarose gel is used. For larger plasmids, 0.3 to 0.7% agarose gels are prepared while small fragments may be visualized on 2% agarose gels.

6. Using a Pasteur pipette, seal the inner edges of the spacers with molten agarose. Let the agarose solidify.

6. This step prevents leakage under the spacers when the gel is subsequently poured.

7. Cast the gel with the remaining agarose, after it has cooled a little. Take care not to introduce bubbles into the gel. Insert the well-forming comb and allow the gel to solidify.

7. Bubbles can be removed by aspiration with a Pasteur pipette or by puncturing them with a hot inoculating needle. The agarose will become opaque after it solidifies.

8. While the gel is solidifying prepare DNA samples for electrophoresis. Add approximately 1 μg of DNA (20−25 μL) to 5−10 μL of buffer containing 30% sucrose and 0.01% bromophenol blue.

8. ***DNA SAMPLE PREPARATION:*** to aid loading, the sample solution must be converted to a density greater than that of the electrophoresis buffer. This is achieved by mixing the appropriate amount of DNA with sucrose, ficoll, or glycerol to give a final concentration of 6−10% for the first two chemicals and 3−9%

for glycerol. To assist in monitoring the progress of the separation, the tracking dye, bromophenol blue, which migrates more rapidly than most small DNAs, is also added to the sample solution. Other tracking dyes, such as xylene cyanol (0.015%) may also be used in combination with bromophenol blue. Xylene cyanol runs with larger size DNA molecules.

9. After the agarose has solidified, remove the comb with a gentle back and forth motion taking care not to tear the gel at the bottom of the well.

MARKER DNA: In analyzing circular DNA molecules, one must remember that both the buffer and the voltage will affect the migration of these DNAs. Therefore, the marker DNA must also be circular, not linear (i.e., do not use restriction digests as marker DNA in this case).

10. Remove the end spacers and place the gel, supported by the base plate on the central supporting block of the buffer chamber (Figure 1.1*b*). Position the sample wells at the cathode end (black).

11. Immerse the gel in the appropriate 1× electrophoresis buffer. Add buffer until it reaches a level approximately 3−5 mm above the surface of the gel.

11. The gel can be either submerged in electrophoresis buffer (submarine gel), or absorbent wicks can be dipped into the buffer chambers and draped over the ends of the gel (dry gel).

If the apparatus in Figure 1.1*a* is used, about 800 mL of buffer is needed.

12. Connect the electrophoresis chamber to the power source. **Do not turn the power on.**

12. Connecting the electrodes after loading the samples might disturb the samples in the wells.

13. Load the sample mixtures into the wells using a plastic-tipped micropipettor. Take care not to overload the well or to "spill" the sample into a neighboring well. Record the well and sample numbers.

13. CAUTION: *Make sure the power is off while loading.*

14. Cover the apparatus with a safety cover and turn on the power. Adjust the voltage to 75V (this usually provides a current of 150 mA). Run, without cooling, for 3 hours.

Electrophoresis is generally continued until the bromophenol blue tracking dye has run off the gel.

14. *POWER SUPPLY:* for separating nucleic acids on agarose or polyacrylamide gels, power supplies that can deliver up to 500 volts and 200 mA are adequate. For DNA sequencing, power supplies capable of delivering up to 2000 volts are required. Therefore, it might be economical to obtain a power supply that can be used for both purposes.

15. After electrophoresis is complete, turn off the power supply and disconnect the cables. Remove the gel-supporting base plate and slide the gel into a solution of ethidium bromide to stain the DNA. Take care not to tear the gel.

15. See Section VI, p. 28, for staining protocols.

16. Destain and photograph the gel as outlined in Sections VI, p. 28 and VII, p. 29.

17. If the gel is to be retained for other purposes (e.g., Southern blots), it should be wrapped in a plastic film (such as Saran Wrap), labeled and stored at 4°C.

17. It is best to use the gel immediately if DNA is to be transferred to other matrices. Wrapping with plastic film prevents the gel from drying. Keep the gel from freezing so that "crystals" do not form in the gel.

METHOD 2: VERTICAL AGAROSE GELS

1. The apparatus for casting vertical gels is shown in Figure 1.2*a*.

2. Wash the plates with ethanol to remove grease or debris.

3. Place the side spacers between the front and back plates. Align the edges of the spacers with those of the plates, then clamp into position (Binder Clamp No. 10).

Immobilize the spacers and plates by taping the edges with waterproof tape (such as Scotch brand 3M Tape).

4. Set the gel-casting apparatus in a small Plexiglas trough (see Figure 1.2*a*). Clamp the plates near the bottom to prevent leakage from the sides of the plate when casting the gel.

4. The trough makes it easier to pour the gel and to make an agarose plug. Gels having agarose concentrations under 1%, are generally not rigid enough to be self-supporting and could slide out from between the plates when placed in an upright position. Therefore, a piece of nylon screening material or nylon stocking should be taped over the bottom of the plates for support. Neither interferes with electrophoresis.

5. Add molten agarose along the inside edges of the spacers with a Pasteur pipette such that an agarose plug, 1 cm high, is formed at the bottom of the casting assembly. Allow the agarose plug to solidify.

5. Running the agarose down the edges to form a plug seals the edges and bottom, thereby preventing molten agarose from leaking from the apparatus while casting the gel.

6. Pour the remaining agarose into the casting assembly to within 1 cm of its top. Take care not to introduce air bubbles.

6. The 1 cm space will prevent the molten agarose from overflowing when the comb is inserted.

7. Insert the well-forming comb in such a way as to avoid trapping air under its teeth. Shrinkage away from the comb sometimes occurs as the agarose cools. If so, simply top up the gel with molten agarose. Let the gel solidify.

7. The comb is essentially the same one used for horizontal gels. The comb may be redesigned so that the number and depth of the wells is varied according to the number and volume of samples to be run.

8. Remove the solidified gel from the plug-forming trough.

9. Place the notched side of the gel-forming plates against the notched side of the upper chamber of the vertical gel apparatus (Figure 1.2*a*). Clamp the plates in position.

9. Both sides of the upper chamber are notched so that two gels may be run on the same apparatus. The chamber can be cooled by running water through the portals.

10. Clamp a Plexiglas plate onto the other side of the apparatus.

10. If only one gel is being run, the other notched side of the apparatus must be sealed off so as to form a complete upper buffer chamber.

11. Place the entire apparatus on the central supporting block (buffer chamber from horizontal assembly).

12. To prevent buffer from leaking from the upper reservoir, seal the area between the plates and the chamber apparatus with molten agarose.

12. If leakage occurs after adding buffer, gaps can be sealed by adding molten agarose under the buffer.

13. Add 1× electrophoresis buffer to the lower chamber.

14. After ensuring that the agarose seals in the upper buffer reservoir have solidified, add 1× buffer.

15. Remove the well-forming comb from the gel. To prevent tearing of the gel, first reduce the surface tension between the agarose and the comb by running a small needle down the teeth of the comb. This will physically separate the two components and allow a small volume of buffer to flow between them. Remove the comb using a gentle rocking motion.

15. Tears in the agarose cannot be repaired by mechanically pushing the gel together or by attempting to reseal it with agarose. If the tear runs across several wells, the gel must be recast.

16. Load the DNA samples under the buffer.

16. Prepare samples as outlined in the previous procedure.

17. Connect the electrodes, cover the apparatus and turn on the power.

17. The sample wells in a vertical gel are much deeper than those in a horizontal gel and so are not that easily disturbed. However, connect the electrodes gently. For separating low molecular weight DNA, a low current (15 mA) for approximately 16 hours is desirable. For high molecular weight DNA (14 megadaltons and up) running at a higher current for shorter times gives better results (100 mA, 3 hours).

18. At the end of the run, turn off the power and disconnect the cables.

18. The marker dye (bromophenol blue) should have migrated at least three-quarters of the length of the gel.

19. Carefully remove the gel from the front and back-plates.

20. Stain and photograph the gel as described in Sections VI, p. 28 and VII, p. 29 respectively.

20. A variety of technical problems may be encountered while running agarose gels. These are summarized in Table 1.2.

TABLE 1.2   Troubleshooting Agarose Gels

Problem	Resolution
1. Smiles	May be due to excess salt in the sample (e.g., NaCl, CsCl), overheating or the use of sucrose or glycerol to increase the density of the sample. Ficoll is reported to reduce smiling effects (Southern, 1979).
2. Dye band stays in well	Depends on the method of DNA preparation and the purity of the DNA. Some tracking dyes (e.g., bromophenol blue) decompose in alkali; therefore alkaline-detergent extraction procedures may give this result.
3. Samples do not comigrate	May be due to overheating at high voltages. Gel should be cooled down.
4. Poor separation	Either reduce or increase the gel concentration, depending on the size of the DNA. Lower the concentration to separate large DNAs; increase to separate small DNA.
5. Smearing (especially with rapid isolation method)	May indicate sample overload, which may be resolved by running a slower gel (i.e., overnight separation). It may indicate nuclease contamination or the necessity for treating with RNase.
6. Background fluorescence	This is often a particular problem when visualizing small plasmids and may be due to the presence of contaminating RNA. Run the tracking dye off the gel (RNA runs before bromophenol blue).
7. DNA band lost	The tracking dye can absorb the fluorescence from DNA-bound ethidium bromide. Therefore, faint DNA bands comigrating with the dye may be missed. Samples can be loaded without dye.
8. DNA band migrates differently	A change in electrophoresis buffer or the interaction of DNA with unrelated batches of agarose may produce different migration patterns.
9. Running time too fast or too slow	If the current is too high or the buffer too diluted, the running time will be fast and poor resolution of DNA bands will be achieved. Conversely, a low current or concentrated buffers may lengthen running times. Check concentration of buffers and adjust current.

METHOD 3: POLYACRYLAMIDE GEL ELECTROPHORESIS

A polyacrylamide gel is formed by the vinyl polymerization of acrylamide monomers into long polyacrylamide chains and the cross-linking of the chains with a cross-linking agent, usually N,N'-methylene-bisacrylamide (Bis), in the presence of a catalyst (i.e., ammonium persulfate) and an initiator (i.e., N,N,N',N'-tetramethyl-ethylenediamine-TEMED). The polymerization reaction produces random chains of polyacrylamide, incorporating a small proportion of Bis molecules. These interact with other chains to form a three-dimensional network. The connection of acrylamide used determines the average polymer chain length, while the concentration of Bis determines the extent of cross-linking. Thus, the concentration of acrylamide and its ratio to Bis are important in determining such physical properties of the gel as density, elasticity, mechanical strength, and pore size. By adjusting the total concentration of acrylamide and/or the ratio of acrylamide to Bis, the pore size and thus the sieving effect of the gel can be varied over a wide range. Hence, nucleic acids of different molecular sizes can be separated.

PROCEDURE

1. Prepare the vertical gel casting apparatus as described in Method 2 of this section, p. 22. Wash the plates with distilled water, rinse with ethanol, and position the side spacers (1.5 mm thick) between the plates along the vertical edges. Align their edges and tape them in position with water-resistant tape (cover the entire length of the plate). Also tape the bottom of the plates closed.

2. Prepare a 5% polyacrylamide gel solution according to the scheme in Table 1.3.

TABLE 1.3 Preparation of Polyacrylamide Gels

Stock Solution (in mL)	Concentration of Gel			
	5%	8%	12%	20%
A	6.7	10.7	16.0	26.7
B	28.5	24.5	19.2	8.5
C	0.8	0.8	0.8	0.8
D	4.0	4.0	4.0	4.0

Total final volume is 40 mL.

3. Deaerate the solution under vacuum.

4. Add 15 µL TEMED to the deaerated polyacrylamide solution. Mix.

5. Hold the gel casting apparatus at a slight angle to the bench top and pour in the acrylamide solution. Fill to within 1 cm of the top of the notched plate. Keep the rest of the solution on ice.

6. Quickly insert the well forming comb. Top up with the 5% acrylamide solution.

COMMENTS

1. If a thicker gel is required, use 3 mm spacers. The bottom and edges of the gel casting plates should be sealed (plugged) with 1% agarose or agar melted in electrophoresis buffer.

Some workers prefer using glass plates for this procedure since Plexiglas may bend due to the heat generated during electrophoresis.

2. *CAUTION:* *Acrylamide is neurotoxic and gloves should be worn when handling powders and solutions. After polymerization, it is considered to be nontoxic.*

Stock Solutions for Polyacrylamide Gels

A 30% acrylamide (28.6 g of acrylamide, 1.4 g of Bisacrylamide). Store at 4°C.

B Double distilled water.

C 3% ammonium persulfate (freshly prepared).

D 10× concentrated electrophoresis buffer (either Tris-borate or Tris-acetate).

3. The solution is deaerated when bubbling stops.

7. Position the plates so that they are nearly horizontal. Allow the gel to polymerize at room temperature (this step may take up to 90 minutes).

8. Remove the tape from the bottom of the casting plates and clamp the casting set-up to the upper buffer chamber with large paper clamps.

9. Seal the area between the notched glass plate and the apparatus with molten 1% agar or agarose.

10. Place the upper buffer chamber on the central support (and lower buffer chamber) as indicated in Figure 1.2b.

11. Prepare the samples as described for agarose gels.

12. Fill the chambers with 1× electrophoresis buffer, and gently remove the comb. Load the samples under the buffer as described for vertical agarose gels.

13. Connect the apparatus to a power source and electrophorese at 50 volts for 3 hours.

14. After disengaging the power cables, remove the gel and stain and photograph as outlined in Sections VI, p. 28 and VII, p. 29.

7. The pressure at the bottom of the plates is directly proportional to the height of the gel. Tilting the plates at a small angle reduces the chance of leakage from the bottom by decreasing the pressure at the bottom.

B. INTERPRETATION OF GEL ELECTROPHORESIS RESULTS

Determination of Molecular Nature of Plasmid DNA: The mobility of a DNA molecule during gel electrophoresis depends on its mass and its molecular configuration. If an extract of a plasmid-containing strain has been prepared, plasmid DNA and chromosomal DNA and RNA will be present in the gel. All three are detected by the staining techniques described in Section VI, p. 28. Due to its large size, the chromosome is sheared into linear fragments during cell lysis and DNA isolation. In most instances, this DNA can be expected to appear within the first third of the gel. Because chromosome fragment sizes differ, the band is usually rather diffuse. However, the chromosomal band from several separate strains or colonies generally migrates the same distance on the gel and is readily identified as a common band. Large DNA (> 320 kb pairs) and protein-DNA complexes remain at, or near, the origin of migration (sample well). Ribosomal RNA moves very quickly and usually co-migrates with the bromophenol blue tracking dye. The remaining bands are generally plasmid DNA.

The covalently-closed circular plasmid form migrates faster than the linear form which, depending on the buffer used, may run faster than the open circular form. All three molecular forms might be found in a single preparation. This can create a problem when analyzing cells containing two or more separate plasmids. Indeed, it may lead to problems in ascertaining exactly how many distinct plasmids are present. This difficulty may be resolved by introducing nicks into covalently-closed circular (CCC) plasmid DNA by uv irradiation (Brunk and Simpson, 1977), thereby converting most of the DNA into the open circular form. The bands displaying an increased ethidium-bromide fluorescence after this treatment would represent open circular forms of the plasmids, while the bands representing the CCC form would decrease in intensity or disappear. In a second approach, the open circular forms may be converted to linear molecules by heating the plasmid preparation to 100°C for 3−7 min in the presence of 0.1% sarkosyl (Van den Hondel *et al.*, 1979). Closed circular forms are unaffected by heat and show no change in band intensity, while open circular forms are reduced in intensity or eliminated. Linear molecules would appear, or increase in intensity.

Determination of Molecular Weight: The simplest way to determine the molecular weight of a DNA molecule is to compare its relative mobility (Rm) on an electrophoresis gel with those of molecules of known molecular weights. The relative mobility of a molecule is the ratio of distance migrated by the unknown molecule to that of a marker substance, usually the tracking dye or the largest molecular weight reference standard (Meyers *et al.*, 1976). With slab gels, the migration distance of the tracking dye is presumably constant throughout the gel, therefore the relative mobilities of molecules running on the same gel can be taken as the distance migrated by the molecules. The distance of migration (or Rm) for each marker molecule from the origin (bottom or edge of the well) is measured. This value is plotted against the log of the molecular weight (MW) of the marker (this can be done directly on semi-logarithmic graph paper). The migration distance of the unknown molecule is measured, and its molecular weight is extrapolated from the plot. Such an Rm vs log MW plot usually produces a curved line; therefore molecular weights obtained from the top or the bottom of the curve will not be accurate.

In constructing the standard plot, it is always advisable to use markers with molecular weights close to that of the DNA molecule being investigated. Molecular weight markers for linear DNA can be obtained commercially or by digesting well-studied plasmid or viral DNA with restriction endonucleases. For example, a *Hind*III digest of λc1857Sam7 DNA gives 8 fragments with molecular weights of 23.6, 9.64, 6.64, 4.34, 2.26, 1.98, 0.56, and 0.14 kilobase pairs (Philippsen *et al.*, 1978). As mentioned before, the conformation of a nucleic acid influences its mobility, therefore circular DNA markers should always be used for estimating the molecular weight of circular plasmid DNA. Any well-studied plasmids such as pBR322 (4.3 kbp), *Col*E1 (6.4 kbp), pMB9 (5.45 kbp), RP4 (54.5 kbp) and R1drd19 (95.45 kbp) may serve as circular DNA markers for molecular weights (see Bukhari *et al.*, 1977).

C. GEL ELECTROPHORESIS IN TWO DIMENSIONS

Electrophoresis in two dimensions (2-D) was initially developed for the separation of mixtures of proteins (see Garrels, 1983, for a recent summary). After some modification, the technique is now being applied in the separation of complex mixtures of nucleic acids (for recent examples, see Hu *et al.*, 1983 (RNA) and Gilroy and Thomas, 1983 (DNA)).

Briefly, a mixture of nucleic acids is electrophoresed in one direction (e.g., down a vertical gel) and is then rotated 90° and subjected to a second electrophoretic separation. In practice, the region of interest is physically removed (cut out) from the first gel, rotated, and then incorporated into a second gel for continued separation. The resultant pattern will have a more or less diagonal shape. In an ideal system, the relative order of migration of each member of the mixture would be very different in the two dimensions. For RNA separations, such alterations in migration rates are obtained by varying the gel concentration or pH, and by the inclusion of a denaturing agent such as urea. Different distribution of DNA mobilities is achieved using gels of different compositions in the two dimensions (agarose versus polyacrylamide), urea gradients, and elevated temperatures (50−60°C). The latter effects a partial denaturation of double-stranded DNA, a condition essential for successful separation. In both instances the parameter altered is determined, in part, by the size and the number of the constituents of the nucleic acid mixture. More detailed information on this type of 2-D gel electrophoresis may be obtained from DeWachter and Fiers (1982) and Fischer and Lerman (1979a).

Two dimensional electrophoresis has been used in the analysis of messenger-RNAs (Burckhardt and Birnstiel, 1978), transfer RNAs, 5S and 6S RNAs (Ikemura and Dahlberg, 1973) and viral RNAs (DeWachter and Fiers, 1972). It has been used less extensively with DNA, but good separations of *E. coli* and λ restriction endonuclease fragments (Fischer and Lerman, 1979a) and *Drosophila* repetitive DNA (Smith and Thomas, 1981) have been reported. By comparing the separation patterns obtained with λ-lysogens and nonlysogens (i.e., *E. coli* chromosomal DNA), Fischer and Lerman (1979b) were able to detect DNA fragments unique to lysogenic *E. coli*. A number of these were believed to be composed of both viral and bacterial DNA. This observation, coupled with the ability to identify 350 of 470−610 postulated restriction endonuclease *Eco*RI-generated fragments, suggests that 2-D electrophoresis may find application as a method for the direct selection of recombinant DNA (see Section X, p. 40).

D. AGAROSE-POLYACRYLAMIDE COMPOSITE GELS

Polyacrylamide-agarose composite gels are primarily used for separating high molecular weight RNA. Because the low-concentration polyacrylamide gels required to separate high molecular weight RNA (<3%) are very fragile and difficult to handle, they are of little use in separating molecules with a molecular weights above 1.6 kilobases (kb). To overcome the difficulty in handling fragile gels, Uriel and Berges (1966) used composite gels of agarose and polyacrylamide for protein separations. Composite gels were first used for the separation of nucleic acids by Peacock and Dingman (1968). In composite gels, different concentrations of agarose (most often 0.5%) have been used to strengthen the low concentration of polyacrylamide (<3%). The inclusion of agarose does not contribute to the separation of molecules.

Composite gels are prepared either by keeping the buffer containing agarose and all the polyacrylamide reagents above the gelling temperature of agarose (35°C) until the polymerization of acrylamide has finished, or by allowing the agarose to solidify first at room temperature (20−25°C) before initiating the polymerization of acrylamide. Both procedures produce satisfactory results (Uriel and Berges, 1966; Peacock and Dingman, 1968). The electrophoresis and buffering conditions for composite gels are the same as those for polyacrylamide gels.

VI. Staining DNA with Ethidium Bromide

DNA bands can be visualized in agarose and polyacrylamide gels by staining them with the intercalating dye ethidium bromide, which fluoresces under ultraviolet illumination. The method is very sensitive and as little as 10 ng of DNA is readily seen. It should be remembered that single-stranded nucleic acids are also stained with this procedure, although to a much lesser extent. The two methods commonly used to stain DNA with ethidium bromide are listed below and are modifications of the procedures of Aaij and Borst (1972) and Sharp *et al.* (1973).

METHOD 1: STAINING GELS AFTER ELECTROPHORESIS

PROCEDURE

1. Prepare a **STOCK SOLUTION** of **ETHIDIUM BROMIDE** by dissolving 0.1 g of ethidium bromide in 10 mL of water (final concentration 10 mg/mL). Store the stock solution in a brown bottle.

2. Add 1 drop (20−25 µL) of the ethidium bromide stock solution to 250 mL of water (final concentration 0.5−1.0 µg/mL). Mix well and pour into a staining tray.

3. Slide the slab gel into the tray and stain for 15 minutes in the dark.

4. Pour off the staining solution and destain the gel for another 15 minutes in an excess of tap water.

5. Visualize DNA bands as described in Section VII, p. 29.

COMMENTS

1. *CAUTION: Ethidium bromide is mutagenic. Vinyl gloves should be worn at all times.*

The use of ethidium bromide for staining DNA is the method of choice. However, a variety of other dyes such as Stains-All, acridine orange, pyronine B, methyl green, toluidine blue O, etc. have been used to stain both DNA and RNA (see Gaal *et al.*, 1980; Andrews, 1981). In addition, the silver stain developed by Merril *et al.*, (1979) has been used to stain DNA and RNA on polyacrylamide gels and is reputed to be 2−5 times more sensitive than ethidium bromide (BioRad Laboratories). It may be the preferred agent for staining denaturing gels as the ethidium bromide color fades quickly in these gels.

2. The volume of staining solution used varies with the size of the staining tray. The best volume is the one that completely covers the gel.

3. The gel, particularly a "sloppy" one, may be more easily manipulated if supported by a piece of nylon screening.

The staining time depends on the concentration of the gel. For gels with agarose concentrations under 1%, 15 minutes is generally sufficient. For gel concentrations higher than 1%, 30−60 minutes might be required.

4. The destaining step may be omitted if the level of background fluorescence is not critical.

5. RNA present in certain DNA preparations may give rise to a general background haziness. Ribosomal RNA, which comigrates with the bromophenol blue tracking dye, may obscure small DNAs. RNA may be eliminated by treatment of cell lysates with RNase; however, it must be remembered that most RNase preparations are contaminated with DNase activities (see Section XVII-G, p. 99). Destaining also eliminates background fluorescence.

METHOD 2: INCORPORATION OF ETHIDIUM BROMIDE INTO THE GEL

PROCEDURE

1. Add 50 µL of the ethidium bromide stock solution (Method 1) to 1 litre of 1× electrophoresis buffer (final concentration 0.5 µg/mL).

COMMENTS

2. Prepare the slab gel (see Section V, p. 18) with ethidium bromide-supplemented electrophoresis buffer.

3. Electrophorese the DNA sample using the electrophoresis buffer prepared in step 1.

3. The progress of the DNA can be followed during electrophoresis by occasionally shining a longwave uv light source over the gel. Alternatively, the gel may be placed on a transilluminator, providing that the Plexiglas used to support the gel is uv permeable.

4. When electrophoresis is complete, visualize the DNA as described in Section VII, p. 29.

4. There are several possible disadvantages to this method.

a. The intercalation of EtBr may alter the conformation of a particular DNA molecule thereby changing its mobility.

b. Incandescent light produces nicks in DNA preparations containing ethidium bromide (Martens and Clayton, 1977).

VII. Visualization and Photography of Stained DNA

The fluorescence of ethidium bromide-DNA complexes is optimally excited by wavelengths of 300 nm (Brunk and Simpson, 1977). The resultant fluorescence emission is maximum at 590 nm. Several commercially available uv light sources have outputs of 254, 300 and 366 nm. In comparing these light sources, Brunk and Simpson (1977) found that while illumination at 254 and 300 nm produced highly fluorescent DNA bands as compared to the 366 nm source, the 254 nm light source produced rapid photobleaching (ethidium bromide is dissociated from the DNA) as well as the highest damage to the DNA in terms of nicking and dimerization. Although the 300 nm light source produces some nicking, it is felt that the optimum excitation spectrum offsets this disadvantage.

Ethidium bromide-DNA complexes in gels may be illuminated either from the sides (incident light) or from the bottom (transmitted light). The transillumination system is the method of choice since faint bands may not be seen with incident light. However, uv transilluminators are expensive, making incident uv illumination more attractive economically.

METHOD 1: INCIDENT UV ILLUMINATION

| PROCEDURE | COMMENTS |

1. Use unfiltered, standard germicidal lamps (with two GE G15T8 15-watt germicidal tubes).

1. *CAUTION: Safety goggles should always be worn around uv light sources.*

2. Place the ethidium bromide-stained gel on a black background and illuminate with two germicidal lamps fixed at an angle of 10−20 degrees above each side of the gel.

2. Since they tear easily, the manipulation of slab gels containing less than 1% agarose is difficult. A piece of nylon window screen may be used to support the gel while transporting it from staining solution to the photography table.

3. Photograph the gel with a Polaroid MP4 camera which is placed directly above the gel. A Kodak Wratten

3. Polaroid type 52 film (positive) or type 55 film (positive/negative) is used with a 545 film holder, while

22A or 23A red filter is used on the camera to remove red light emitted by the uv lamp.

type 667 film is used with a 44-48 holder adapted for the MP4 camera. For incident uv illumination, type 52 film is exposed for 1−2 minutes. If type 55 film is used, exposure time is 5−8 times longer. It should be noted that faint bands, not visible to the naked eye may appear if a negative is prepared. It should be noted that faster films may be used, although they are generally more expensive.

A standard 35 mm camera may also be used. Although it will give better picture resolution, it is not "instant" and the film requires photographic development.

METHOD 2:   TRANSILLUMINATION

1. Support the ethidium bromide-stained gel above the uv (300 nm) light source with a uv pass-visible block filter. Ultraviolet light passing through the gel causes the stain to fluoresce.

1. The uv pass-visible block filter (available from Corning) also known as an excitation filter, blocks light in the visible wavelengths and permits light in the 300 nm range to pass at 70% efficiency (Brunk and Simpson, 1977). Since many of the 300 nm lamps produce wavelengths in the red spectrum which are not blocked by the excitation filter, a red filter must be placed on the camera. Commercially available transilluminators supply the appropriate light source (not always 300 nm) and excitation filter.

2. The camera set-up is the same as for incident uv illumination.

2. The film types are the same as those described with incident uv illumination (e.g., Polaroid type 52 or 55 film). Film exposure times are greatly reduced (e.g., exposure time for type 52 is about 3−5 seconds) by transillumination. However, these times may vary with the age of the transilluminator filter which solarizes in proportion to time used (solarization drastically cuts short wave transmission).

VIII. Recovery of DNA from Slab Gels

Frequently, a particular DNA molecule must be recovered from gels for a variety of purposes (e.g., restriction endonuclease digestion, sequence analysis or transformation). Several methods accomplish this. All are based either on the electroelution or the disruptive release of nucleic acids from the gel matrix. The disruption methods are either physical or chemical and are employed to recover DNA from both horizontal and vertical gels. Electroelution techniques are used primarily with horizontal gels. It is important that the DNA be free of certain chemicals (e.g., agaropectin) present in agarose that contain substances (e.g., sulfur) that may inhibit enzymes such as ligase. This may be achieved by precipitating the recovered DNA several times or by using ultrapure agarose.

METHOD 1: ELECTROELUTION ONTO DIETHYLAMINOETHYL (DEAE) PAPER

PROCEDURE

1. Visualize the ethidium bromide-stained DNA band using a long wave uv light source.

2. Cut a slit (slightly longer than the band) just in front of the DNA to be recovered. Replace the gel on the base plate of the horizontal apparatus.

3. Cut a piece of DEAE paper that is 0.5 cm wide and as long as the slit in the gel.

4. Wet the DEAE paper with electrophoresis buffer (1×). Using forceps, insert it into the slit in the gel.

5. Reposition the base plate on the electrophoresis apparatus maintaining the gel's original orientation, i.e., sample wells at cathode. Add electrophoresis buffer to the reservoirs so that it is level with the upper surface of the gel (don't cover it). Fill the sample wells with buffer.

6. Electrophorese for 5−15 minutes at 100 mA.

7. Remove the DEAE paper and determine whether the DNA band has moved from the gel onto the paper by longwave (366 nm) uv irradiation.

8. Place the DEAE paper in a 500 μL microfuge tube and add 200 μL of Tris-acetate buffer (i.e., 1× electrophoresis buffer with pH adjusted to 8.0 with glacial acetic acid) containing 1N NaCl.

9. Let the tube stand at room temperature for 2−3 hours, or overnight at 4°C.

10. Punch a hole in the bottom of the tube with a small needle and place it into a 1.5 mL microfuge tube. Centrifuge for 1 minute at full speed.

11. Precipitate the DNA by adding sodium acetate to a final concentration of 0.3M and then 2 volumes of cold 95% ethanol. If the DNA is to be used for transformation, carrier transfer RNA may be added (final concentration 0.1 mg/mL) to increase the yield of precipitate.

COMMENTS

1. The following method is adapted from Banner (1982) and Winberg and Hammarskjöld (1980).

2. See Figure 1.1 for a description of horizontal gel electrophoresis apparatus.

3. DEAE is an ion exchanger and binds nucleic acids at low salt concentrations (low ionic strength) and releases them at high salt concentrations (high ionic strength).

4. Use the same buffer as was used for electrophoresis. Make sure that the entire cut-edge of the gel is in contact with the paper.

5. Buffer is placed in the sample wells to maintain electrical continuity.

6. Longer times (30−45 mintues) may be required to achieve the elution of larger DNAs.

7. If the DNA band has not migrated into the DEAE paper, repeat step 6.

8. Alternatively, one could cut the DEAE paper into small pieces before adding the elution buffer.

10. If the DEAE paper was cut into small pieces the following protocol should be used. Punch a hole in the bottom of a second 500 μL microfuge tube and place a small wad of siliconized glass wool at its bottom. Place this tube inside a 1.5 mL microfuge tube. Transfer the DNA solution (elutant) to the smaller tube and centrifuge. The DEAE paper will be retained by the glass wool.

11. If the DNA is to be used for restriction endonuclease analysis or for transformation, ethidium bromide must be extracted with isobutanol before precipitation (Section IV, p. 12).

PROCEDURE	COMMENTS

12. Resuspend the DNA in the appropriate buffer.

METHOD 2: ELECTROELUTION INTO A BUFFER TROUGH

1. Locate the DNA band of interest in the ethidium bromide-stained gel using a long wave uv light source.

1. This procedure has been adapted from Yang *et al.*, 1979.

2. Cut and remove a small fragment of the gel such that a small trough, approximately the same size as the band itself, is formed immediately in front of the band (anode side).

3. Return the gel to the base plate of the horizontal apparatus. Fill the trough and the buffer reservoir with 1× electrophoresis buffer. Do not cover the gel with buffer.

3. The trough will usually hold 30 μL.

4. Connect the apparatus to a power source and apply the same current as used in the initial separating run.

5. Monitor the migration of the band into the well with a hand-held longwave uv source. The time required for elution depends on the size of the DNA in question. When eluting small molecules at high voltage (74–100 V), check the progress every 2 minutes.

5. Alternatively, run the sample in duplicate. Use one for recovery and the other for monitoring.

6. Pipette the buffer from the trough into a microfuge tube and adjust its volume to at least 100 μL with fresh buffer.

6. When recovering high molecular weight molecules (23–38 kb) several changes of trough buffer may be made.

7. Extract the ethidium bromide (Section IV, p. 12). Adjust the 0.3M sodium acetate and precipitate the DNA with 2 volumes of cold ethanol. (See Section I, p. 4.)

8. Resuspend the DNA in an appropriate buffer.

8. Duplex DNA up to 37 kb can be recovered with reasonable efficiency.

METHOD 3: "OPTIMIZED" FREEZE-SQUEEZE PROCEDURE

This procedure, originally developed by Thuring *et al.* (1975) and further adapted from the modifications made by Tautz and Renz (1983), provides a rapid and simple means for the recovery of DNA from agarose gels. Circular forms up to 32 kb in mass and 50 kb linear molecules have been isolated with varying degrees of efficiency. The conversion of a number of closed circular molecules to open circular forms may occur during recovery by this method.

PROCEDURE	COMMENTS

1. Locate the ethidium bromide-stained band of interest using a long-wave uv light source.

1. To minimize the conversion of ethidium bromide-intercalated supercoils to open-circular forms, as well

PROCEDURE	**COMMENTS**

as to eliminate the need to remove ethidium bromide at subsequent steps, the DNA sample may be divided into two aliquots: one with a small volume for staining (detection) and a second of large volume for recovery. Electrophorese in adjacent lanes. Immediately prior to staining cut away the portion of the gel containing the larger amount of DNA and store separately. Stain the remaining portion of the gel. Locate the DNA band and mark its position in the gel (e.g., with a wooden toothpick). Remove the uv light, then align the stained and unstained portions of the gel to determine the location of the ethidium bromide-free DNA.

2. Cut out the portion of the gel containing the desired DNA and place it in a 0.6 mL microfuge tube. The bottom of the tube should hold a small wad of siliconized glass wool.

2. Avoid removing extraneous gelling material, as best recoveries are achieved with small slices. The gel may be cut into smaller pieces to improve recovery, but it must be remembered that such manipulations may increase the concentration of open circular DNA. The preparation of siliconized glass wool is described in Section XVII-H, p. 100.

3. Add 50–150 μL of elution buffer #1 to the tube.

3. Elution Buffer #1

0.3M sodium acetate
1 mM EDTA
pH 7.0

A volume of 100 μL is used routinely for slices averaging $6 \times 1.5–2.0$ mm in size.

4. Store in the dark at room temperature for 20–30 minutes, then either freeze in liquid nitrogen or at −70°C for 15–20 minutes.

4. Equilibration in high-ionic-strength solution is essential for good recovery. Freezing must be rapid; the use of −20°C refrigerators gives poor yields.

5. Remove the tubes from the freezer and quickly puncture them at the bottom using an 18 gauge needle.

5. Do not allow the agarose to thaw.

6. Immediately place the punctured tubes in 1.5 mL microfuge tubes from which the lids have been removed and spin at maximum speed in a microfuge for 10 minutes at room temperature.

6. Gel debris is retained by the glass wool filter while the elution solution and the DNA passes into the larger tube.

7. Add 1/100 volume of 10% acetic acid–1M MgCl$_2$ solution to the filtrate and mix.

7. Acetate salts increase yields of DNA (see Rapid Isolation Procedures in Section I, this chapter).

8. Precipitate the recovered DNA by adding 2.5 volumes of cold ethanol and incubate at −70°C for 10–15 minutes.

8. The efficiency of recovery is influenced, in part, by the pore size of the gel matrix: best yields are achieved with low concentrations (0.6–0.75%) of agarose.

9. Pellet the DNA by centrifuging at top speed for 10 minutes.

9. DNA recovered in this manner is transformable and serves as a suitable substrate for most restriction endonucleases and other enzymes.

METHOD 4: AMMONIUM ACETATE PROCEDURE

The following method is adapted from Maxam and Gilbert (1977) and Yang *et al.* (1979). It can be used to recover small DNAs (<1kb) from 3.5–5% acrylamide gels, as well as larger molecules from agarose gels. "Contaminants" ionically-bound to DNA may interfere with enzyme activity, particularly that of restriction endonucleases. The ammonium ion of ammonium acetate will dissociate these from the nucleic acid, while the acetate group will convert them to alcohol-soluble salts, thus allowing for the recovery of pure DNA.

PROCEDURE

1. Locate the ethidium bromide-stained band of interest using a long-wave uv light.

2. Cut out the portion of the gel containing this DNA and place it in a small plastic bag. Heat-seal the ends after squeezing out as much air as possible.

3. Place the bag on a flat surface; then thoroughly squash the gel, using the bottom of a glass beaker.

4. Cut open the bag at a corner and add 2mL of elution buffer #2. Reseal the bag and mix the contents. Incubate overnight at 37°C.

5. Using an 18 gauge needle, puncture the bottom of a 1.5 mL microfuge tube. Place a small plug of siliconized glass wool inside. (Section XVII-H, p. 100) Insert the tube partway through a rubber stopper.

6. Attach a plastic test tube (13 × 100 mm) beneath the microfuge tube using masking tape. Place the set-up in a side-arm flask (Figure 1.5).

7. Open the plastic bag and pour its contents into the filter-tube prepared in steps 5 and 6. Turn on the suction.

8. Extract ethidium bromide from the solution using one of the procedures described in Section IV, p. 12. Add a final concentration of 0.3M sodium acetate and precipitate the DNA by adding 2 volumes of cold ethanol as described in this chapter, Section I, Method 1B, p. 4.

COMMENTS

2. A 6–10 cm length of 5 cm wide plastic tubing (3 mm gauge) can be used.

4. Elution Buffer #2

0.5M ammonium acetate
0.01M $MgCl_2$
0.1 mM EDTA
0.1% SDS

7. The elution buffer separates from the agarose as it is drawn through the glass wool plug.

8. DNA recovered in this manner can be used for restriction endonuclease analyses and transformation procedures.

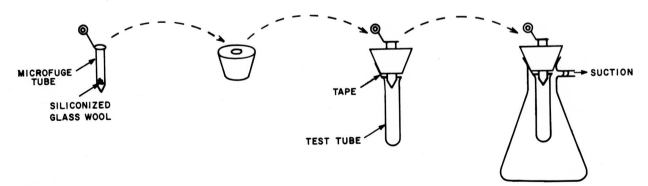

MICROFUGE TUBE

SILICONIZED GLASS WOOL

TAPE

TEST TUBE

SUCTION

FIGURE 1.5 Apparatus for recovering DNA.

Other methods for extracting DNA from slab gels are listed below. In general, the larger the nucleic acid fragment, the less efficient the recovery will be. All of these techniques can be used for recovering DNA from agarose gels; technique 2b can be used for recovering DNA from polyacrylamide gels as well.

1. Disruption Techniques   **References**

   (a) Agarase   Finkelstein and Rownd (1978)

   (b) Chaotropic agents (KI, NaClO$_4$)   Blin *et al.* (1975)
                                                                Vogelstein and Gillespie (1979)

   (c) Thermal   Weislander (1979)

2. Electroelution Techniques

   (a) Hydroxylapatite trough   Tabak and Flavell (1978)

   (b) Dialysis bag   Yang *et al.* (1979)

IX. Restriction Endonuclease Digestion and Mapping of DNA

Restriction endonucleases are enzymes that cleave both strands of DNA at sites internal to the molecule after recognizing specific nucleotide sequences on the molecule. Three types of restriction endonucleases have been described (for reviews, see Boyer, 1971; Arber, 1974; Nathans and Smith, 1975; Yuan, 1981). Type I restriction endonucleases recognize a specific sequence but cleave the DNA randomly some distance away from the recognition site. This type of endonuclease requires ATP, Mg^{++}, and *S*-adenosylmethionine as cofactors. Type II restriction endonucleases recognize a specific sequence and cleave the DNA at a specific site, usually within the recognition sequence. They require only Mg^{++} for their activity. Type III restriction endonucleases require ATP and Mg^{++} for their activities (*S*-adenosylmethionine stimulates the reaction), but cleave DNA 24–26 bases from the 3' side of the recognition site.

Because of their specificity, type II restriction endonucleases have been important tools for the analysis, mapping, and cloning of DNA from both eukaryotic and prokaryotic sources (see Nathans and Smith, 1975; Roberts, 1976; Mercola and Cline, 1980). At present, 355 type II enzymes have been identified in a variety of bacterial strains (Roberts, 1982). These endonucleases recognize specific palindromic sequences (often 4 to 6 base pairs) and cleave the DNA molecule at specific sites within the sequence. The asymmetric digestion of DNA within the palindromic sequence generates tails of DNA that may base pair. For example, the enzyme *Eco*RI recognizes the sequence

$$\downarrow$$
$$5'\text{G-A-A-T-T-C-}3'$$
$$3'\text{G-T-T-A-A-G-}5'$$
$$\uparrow$$

and cleaves the DNA in both strands at sites internal to the recognition sequence. (The arrows indicate the site of cleavage on each strand). The staggered cleavage generates two equimolar molecules with 5' tails.

G		A-A-T-T-C
C-T-T-A-A	and	G

Several different endonucleases generate 5' tails. Other endonucleases such as *Pst*I, which recognizes the sequence C-T-G-C-A$\downarrow$G, generate tails at the 3' end. Both 5' and 3' tails are also known colloquially as cohesive or "sticky" ends. Some enzymes, such as *Hae*III, which recognizes the sequence GG ↓ CC, generate blunt (or flush) ends. A total of 85 different sequences are recognized by the 355 different enzymes. Enzymes that recognize the same sequence are called isochizomers. (Some isochizomers act identically while others recognize identical sequences but cleave the DNA at different sites. Still other isochizomers recognize the same sequence but may not cleave the DNA if it is methylated.) Nomenclature for type II enzymes generally follows the recommendations of Smith and Nathans (1973), in which the producing host is designated by three letters (the first letter is the genus, the next two letters designate species), the strain by a number (e.g., *Agrobacterium tumefaciens* strain B6806 is *Atn*B), and the particular restriction system in the producing organism by a Roman numeral (e.g., *Atn*BI).

The following is a general procedure for digesting DNA with type II restriction endonucleases. For a more comprehensive review of the practical aspects on the use of restriction enzymes, refer to Fuchs and Blakesley (1983).

METHOD 1: GENERAL PROCEDURE FOR RESTRICTION ENDONUCLEASE DIGESTION

PROCEDURE

1. Suspend the DNA in restriction buffer such that its final concentration is $2-3$ µg/mL (minimum volume 50 µL).

COMMENTS

1. Restriction Buffer

100 mM Tris-HCl, pH 7.5
50 mM NaCl
10 mM MgCl$_2$
1 mM dithiothreitol
or
10 mM β-mercaptoethanol

Restriction enzymes vary in their requirements for optimal activity (see Table 1.4). The above buffer is a good, all-purpose buffer for most enzymes including *Eco*RI, *Hin*dIII, *Bam*HI etc. However some enzymes (*Sma*I, *Taq*I) have special requirements and the buffer, as well as conditions of digestion should be amended according to the literature or the manufacturer's specifications. The reagents for restriction buffers should be pure and free of nucleases. Type II restriction endonucleases require Mg^{2+} as a co-factor. Similarly, the reaction is stimulated by the addition of monovalent cations (Na$^+$). The ionic strength optimal for activity of some enzymes is listed in Table 1.4. The sulfhydryl reagents, β-mercaptoethanol and dithiothreitol, are used to inhibit other nucleases. Occasionally, bovine serum albumin is added to the buffer to protect the enzyme from proteases or other factors during long-term storage (see Fuchs and Blakesley, 1983 for more information on core buffer systems).

The DNA used for restriction analysis must be free of agents, such as EDTA, SDS, phenol, and ethidium bromide which may interfere with or inhibit the activity of the enzyme. In addition, compounds such as ethanol, glycerol, and dimethylsulphoxide (DMSO) could change the dielectric constant of the restriction buffer and should be kept to a minimum (Fuchs and Blakesley, 1983). Antibiotics or dyes which bind DNA, such as actinomycin D and distamycin A, may affect the electrophoretic mobility of restriction fragments (Loucks *et al.*, 1979).

The minimum amount of DNA detectable by uv transillumination is about 10 ng. Therefore, with restriction endonucleases producing six fragments per molecule, about $2-3$ μg/mL of DNA is required in order to visualize the bands. Higher concentrations of DNA for similar digests produce thick bands, may cause streaking, and may mask fragments with similar molecular weights. For endonucleases with > 5 sites, a higher concentration of DNA should be used in order to assure that detectable amounts of each fragment are produced.

Although volumes smaller than 50 μL may be used when screening DNA, a variety of factors (e.g., component concentration, pipetting error, etc.) may introduce error (see Fuchs and Blakesley, 1983). These errors may be critical in digestions undertaken for analytical purposes.

NOTE: *Many workers sterilize (autoclave) all glassware, buffers, and other reagents (when possible) to reduce the possibility of contamination with unwanted DNases.*

TABLE 1.4 Ionic Strength for Optimal Activity of Certain Restriction Enzymes

Low (0–10 mM NaCl)	Medium (30–60 mM NaCl)	High (75 mM NaCl)	KCl
*Bal*I	*Alu*I	*Bam*HI	*Hpa*I (50 mM)
*Bcl*I	*Ava*I	*Bgl*I[b]	*Mbo*II (6 mM)
*Cfo*I	*Ava*II	*Dde*I	*Sma*I (15 mM)
*Cla*I	*Bgl*II[b]	*Hinf*I	
*Hae*II	*Bst*EII[a]	*Mbo*I	
*Hae*III	*Dpn*I	*Sal*I	
*Hind*II	*Eco*RI	*Sau*961	
*Hpa*II	*Eco*RII	*Sph*I	
*Kpn*I	*Hha*I	*Sst*I	
*Msp*I	*Hinc*II	*Sst*II	
*Nci*I	*Hind*III	*Taq*I[c]	
*Tha*I[a]	*Pst*I	*Xho*I	
*Xor*II	*Pvu*II		
	*Sau*3A		
	*Xba*I		

Unless specified, the buffer system used is Tris-HCl (pH 7.5), and the incubation temperature is 37°C.

[a]Incubation temperature is 60°C.
[b]Buffer system is glycine-NaOH.
[c]Incubation temperature is 65°C.

2. For single digestions, add sufficient restriction endonuclease to give a final concentration of 5 units.

2. One unit of restriction endonuclease is defined as the amount of enzyme required to completely digest 1 μg of λ DNA at the appropriate temperature in one hour. Digestion may be influenced by a variety of parameters. For example, the base composition adjacent to an endonuclease recognition sequence can affect the restriction rate by as much as 25-fold (Modrich and Rubin, 1979). Some enzymes (e.g., *Ava*I, *Bam*HI, *Hae*III etc.) show altered recognition specifications when standard reaction conditions are varied (Fuchs and Blakesley, 1983). Furthermore, more enzyme may be required to digest CCC DNA than linear DNA.

3. Incubate the digestion mixture at 37°C for 2–3 hours.

3. These are minimum incubation times for complete digestion. If a longer incubation time is used, a stabilizer such as nuclease-free bovine serum albumin should be added to the reaction mixture. Nonspecific nucleases may cause problems during longer incubation periods.

4. *DOUBLE DIGESTIONS:* Where the buffer conditions are identical, add 5 units of the second enzyme and incubate as above.

4. Sequential digestions (i.e., sequential addition of endonuclease), ensure that each enzyme completely digests the DNA. However, simultaneous digestions may also be conducted if the buffer requirements are identical. If enzymes have different requirements, the general buffer must either be modified accordingly by increasing the molarity, or replaced. The enzyme with the lower ionic strength requirement should be used first.

5. At the end of the incubation time add 20 μL of STOP solution (optional).

5. The ratio of stop solution to restriction mixture should be 1:3.

STOP Solution

 4M urea
50% sucrose
50 mM EDTA
0.1% bromophenol blue
pH 7.0

Urea is added to prevent reannealing of restriction fragments. Although it is not absolutely necessary to add urea, it is advisable to heat the reaction mixture to 65°C (to prevent reannealing of sticky ends), followed by immediate cooling on ice before loading the sample onto a gel. EDTA is added to complex the Mg^{++} ions that are required for endonuclease activity, thus preventing further digestion. Sucrose increases the density of the DNA solution so that it stays beneath the buffer when loading the gel. Bromophenol blue is a tracking dye.

6. Load a slab gel with the entire digested sample and electrophorese.

6. The amount of sample loaded on the gel depends on the concentration of DNA and the number of fragments generated upon digestion with the enzyme used.

It is advisable to run molecular weight markers on the same gel (see Section on DNA mapping, p. 39).

7. Stain the gel with ethidium bromide (Section VI, p. 28) and determine the molecular mass of the fragments.

7. The interpretation of restriction endonuclease digests may not be clear-cut. A variety of problems ranging from the failure to cleave the DNA, the presence of more fragments than expected, diffuse bands, to no

DNA, may be encountered (see Fuchs and Blakesley, 1983 for an excellent review). Solutions to these problems range from "cleaning-up" the DNA preparation to using more enzyme for cleavage.

MODIFICATION of DNA may contribute to anomalous digestion results. A given bacterial host may protect itself from its own restriction endonucleases by modifying (by methylation) the recognition site. The methylated nucleotides are generally 5-methylcytosine or N^6-methyladenine (Ehrlich and Wang, 1981). Thus, one enzyme may cleave unmethylated DNA while another, which recognizes the same sequence, may cleave methylated DNA.

DNA MAPPING

Several techniques have been used to construct restriction endonuclease maps of a DNA molecule. These techniques include single and multiple digestions; secondary digestion with another enzyme following the isolation of an endonuclease-generated fragment; partial digestion of a ^{32}P end-labeled fragment; and digestion with an exonuclease such as *Bal*31 followed by digestion with the enzyme(s) of interest. The technique chosen for mapping is determined by the level of detail required. Very often, a combination of techniques is used (Yoneda *et al.*, 1979).

A series of single digestions with the unknown DNA molecule should be carried out as the first step in mapping. This determines the number of restriction sites on the DNA for each enzyme, and the molecular weight of each fragment generated. One must be careful in determining the number and molecular weights of fragments produced, as tiny fragments may run off the gel, and large fragments may not be resolved in the gel system used. Therefore, it is advisable to analyze the enzyme digest along with the appropriate molecular weight markers. Different gels (i.e., polyacrylamide or agarose) of different concentrations should also be run, to ensure that no fragments are lost. With circular DNA, the number of fragments generated is equal to the number of cleavage sites. With linear DNA, the number of cleavage sites is equal to one less than the number of fragments generated. Denaturing gels are quite often used to accurately determine the molecular weight of a DNA fragment (Section XV, p. 88).

In single and multiple digestion mapping, the restriction sites of one enzyme are oriented with respect to the restriction sites (reference sites) of a "reference enzyme." The best starting reference enzyme is one that generates only 2−3 fragments—that is, 2−3 sites on a circular DNA molecule or 1−2 sites on a linear DNA molecule. The reason for choosing such a reference enzyme is that, except for the problem of mirror images, the restriction sites can be fixed unambiguously with respect to each other. Once fixed reference sites are established, a series of double digestions (multiple digestions) can be performed to determine the restriction sites of other enzymes. Double digestions are performed by either digesting the DNA molecule with both enzymes simultaneously, or sequentially (Jorgensen *et al.*, 1979; Tenover *et al.*, 1980). Fragments that "disappear" after the second digestion contain sites for the second enzyme. New bands appearing represent unique double digestion products. Molecular weights of the fragments are determined and, from the analysis of all the data, a physical map can be constructed.

Partial digestion was first successfully used by Smith and Bernstiel (1976) to create a restriction map of sea urchin histone DNA. In this technique, the DNA molecule must be linear (circular DNA is linearized with a single-site restriction endonuclease), and one of its ends labeled with ^{32}P. To obtain a DNA fragment with only one labeled end, both ends must first be labeled using either 3′ or 5′ end-labeling (Section XIIB, p. 70). The fragment is then digested with a restriction enzyme that has only one site on the DNA molecule. The two fragments so generated have only one of their ends labeled, and can be separated by gel electrophoresis. Partial digestion of the fragment is then carried out with the "mapping" restriction enzyme which should digest the fragment at several sites. Its concentration is adjusted such that the molecule will be cut only once. However, since the fragment possesses multiple sites for the enzyme, other fragments will be cleaved at one of the other sites, thereby generating a mixture of fragments of different size. The fragments are then separated by gel electrophoresis and autoradiographed. Each fragment appearing on the autoradiogram represents a restriction site of the enzyme and the molecular weight of each fragment represents the distance of the site from the ^{32}P-labeled end. In determining the molecular weight of each fragment, ^{32}P-labeled molecular weight markers must be used.

Digestion with an exonuclease such as *Bal*31 provides another method for mapping the specific restriction sites of an enzyme (Legerski *et al.*, 1978) on DNA molecules. *Bal*31 degrades both 3′ and 5′ strands of a linear DNA molecule in a progressive manner and its rate of degradation can be controlled. At different time intervals, the exonuclease activity of *Bal*31 can be stopped by adding EGTA (ethylene glycol-bis (β-aminoethyl ether)-N,N,N′,N′-tetraacetic acid). The sample is then phenol/ether extracted to remove the nuclease, and redigested with the enzyme of interest. Samples are removed at different time intervals and run on a gel along with a control (DNA which has not been treated with *Bal*31 but digested with the enzyme of interest). The order in which various fragments disappear on the gel represents the order of the fragments from the ends of the DNA molecule. The fragments are then correctly ordered.

X. Molecular Cloning in Prokaryotic Cells

Attempts at understanding complex systems are often facilitated by breaking them down into smaller units that can then be examined in isolation. Molecular cloning permits this to be done with genetic systems. Using cloning techniques, a single DNA segment may be isolated from a highly heterogeneous mixture of DNAs and propagated to a high degree of homogeneity (amplification) within a bacterial cell. The DNA to be cloned need not be of bacterial origin, but it must be recombined with a genetic element (a vector) that is capable of autonomous replication within a bacterium. (Similarly, cloning in eukaryotic hosts requires vectors capable of replicating within those hosts.) Since bacterial clones are easily detected and grown, large quantities of a specific DNA can be recovered, purified and used for a variety of procedures such as DNA sequence analysis (Hindley, 1983), restriction mapping, transformation, hybridizations, etc.

In general, the DNA to be cloned (foreign DNA) is first dissected into segments (the size depends on the cloning vector and the subsequent use of the DNA) using either physical forces (shearing) or restriction endonucleases. Secondly, the DNA fragments are mixed *in vitro* in the appropriate ratio with a vector molecule, then joined by means of specific DNA joining enzymes (ligases). The resulting recombinant molecules (hybrid DNA's, chimeras) are then introduced by transfection (transformation) into a suitable host cell. To qualify as a suitable host, the bacterium must permit both the cloned DNA to be expressed, and the vector to increase (amplify) the number of its copies. Progeny cells containing the DNA fragment of interest are selected by a combination of genetic and physical methods.

Finally, selected cells are monitored for the functional expression of the cloned genes. The selection and expression steps may be combined where a phenotype can be detected directly. At this stage, consideration must be given to possible effects the cloned DNA may have on bacterial growth as well as to effects the host may have on the stability of the cloned DNA (sequence rearrangements, insert orientation, etc.).

A. CONTAINMENT: PHYSICAL AND BIOLOGICAL

When designing DNA cloning experiments, consideration must be given to the possible biohazards associated with the creation of new combinations of genes (Gorbach, 1978; Freter, 1978; Abram *et al.*, 1982). To minimize, if not eliminate the potential dangers to humans and to the environment in this regard, several agencies (e.g., World Health Organization, Medical Research Council of Canada, United States National Institutes of Health, Genetic Manipulation Advisory Committee—Britain) have developed specific guidelines for the conduct of recombinant DNA work. These guidelines are intended to prevent the inadvertent exposure to, and dispersal of, recombinant molecules through the implementation of stringent physical and biological containment (barrier) systems. Physical containment encompasses several aspects of laboratory design and practice and includes, among other things, the use of safety cabinets and protective clothing, the control of aerosol formation and restricted access to laboratory areas. Biological containment refers to those measures genetically "built-in" to the recombinant molecule or its propagating host, for the purpose of limiting their chances of survival outside the laboratory, i.e., control of dissemination and infectivity.

Both the physical and biological containment systems comprise different levels. In the National Institutes of Health (NIH)—USA system, which is generally recognized as the "gold standard," levels of physical containment are designated P1 through P4. P4 is the most stringent containment level, P1 the least (see Fed. Regist. 48(106), Appendix G, 1983). Levels of biological containment are designated by the letters HV (host-vector) plus the number 1 or 2 (Fed. Regist. 48(106), Appendix I, 1983). An HV1 system provides a moderate level of containment, HV2 a high level.

As more information regarding the real, rather than imagined, levels of biological hazards has become available (e.g., Levy and Marshall, 1981; Levy *et al.*, 1980; Abram *et al.*, 1982), the tendency has been to shift formerly high containment studies to lower or more moderate levels. For example, experiments involving the cloning of DNA in its natural host (self-cloning, e.g., *E. coli* DNA in *E. coli*) using vectors indigenous to that host, or in hosts which exchange genetic information naturally (e.g., the *Enterobacteriaceae*), are now exempt from current NIH guidelines (Fed. Regist. 48:24556, 1983).

Most universities and research institutes have now established specific committees for the purpose of screening and monitoring recombinant DNA-based studies. These should be consulted concerning up-to-date cloning guidelines.

B. GENERAL CLONING VECTORS

To properly serve as a vector for molecular cloning, a genetic element must meet certain requirements: a) It must be capable of autonomous replication within the host so that isolation and amplification of a specific DNA fragment can be achieved. b) It must possess (nonessential) regions within which the nucleotide chain may be cleaved and foreign DNA inserted without affecting the replication of the vector or the ability to select recombinant molecules. c) It must carry genes that confer an easily recognized phenotype upon the host bacterium so that vector-carrying cells can be selected (e.g., antibiotic or bacteriocin resistance, sugar fermentation). d) It must possess a single cleavage site for at least one restriction endonuclease located outside its essential regions at which foreign DNA can be inserted. (The use of an endonuclease with more than one site may lead to the loss of essential replication functions.) e) It should either allow for the regeneration of the restriction endonuclease cleavage sites during recombination, or carry additional endonuclease sites closely adjacent to the ends of the inserted DNA in order to recover an inserted fragment for further study. f) It should be unable to survive outside the host cell, should display a limited host range and be incapable of transmission to other hosts by conjugation, in order to reduce the biohazard associated with the recombinant molecule.

At present, three types of cloning vectors are used for the isolation, amplification and expression of cloned DNA in prokaryotic cells: plasmids, cosmids and phage (Table 1.5). All three enjoy an extrachromosomal existence, replicating in relative autonomy in the host cell. Plasmids are covalently-closed circles of double-stranded DNA, that have been detected in a variety of bacterial genera (Broda, 1979). The larger plasmids usually occur in a limited number of copies per cell (1−3), whereas smaller ones display larger copy numbers (>10). A property of interest in molecular cloning is the ability of certain small plasmids to continue replicating in the presence of the protein synthesis-inhibiting antibiotic chloramphenicol, thereby increasing their copy numbers still further (Clewell and Helinski, 1972; Betlach *et al.*, 1976; Chang and Cohen, 1978). As well as being chloramphenicol-amplifiable, conjugation-defective and nonmobilizable, the ideal plasmid cloning vector should be of small to moderate size (2−10 kb), since transformation efficiency decreases with increasing plasmid size (Collins and Hohn, 1978). Theoretically, foreign DNA of any size may be cloned using plasmid vectors, but to maintain reasonable frequencies of transformation smaller fragments are used. Plasmid hybrids generally are stable molecules.

A large number of plasmid vectors are currently available (see Bernard and Helinski, 1980 for a review of these), with additions each month. Table 1.5 contains a sampling of these, notably the NIH certified and specialty vectors. Plasmid pBR322 (Figure 1.6), is a commonly used cloning vector, (Bolivar *et al.*, 1977a & b), which has been certified as an EK2 plasmid (Fed. Regist. 49:24556, 1983). It is a small (4.3 kb), chloramphenicol-amplifiable plasmid, that determines resistance to ampicillin and tetracycline. Within these two easily-selectable genes, are unique cleavage sites for several restriction endonucleases. Single *Pvu*I and *Pst*I sites are located within the ampicillin gene, while those for *Bam*HI, *Hind*III and *Sal*I lie within the tetracycline gene. Thus, by using one of these sites for the insertion of foreign DNA, recombinant molecules can be distinguished by monitoring for a change in resistance pattern to either ampicillin or tetracycline sensitivity (**insertional inactivation**).

COSMIDS are plasmids that contain the cohesive end sites *(cos)* of bacteriophage lambda. The *cos* region is essential for the packaging of λ DNA into its capsid. Thus, provided the *cos* regions are located in parallel in the DNA and are 37 to 52 kb apart, these vectors can be incorporated into and propagated as complete λ particles (Collins and Hohn, 1978). Since vector molecules and small hybrids are not packaged into transducing particles, cosmids enrich for the isolation of large DNA fragments and are ideal for studying large genes or genetic linkage groups. Restrictions imposed by capsid size and the retention of essential regions limit the size of clonable DNA to 30−45 kb (Hohn and Hinnen, 1980). In addition, hybrid cosmids are not as stable as hybrid plasmids. Cosmid pHC79 (see Figure 1.7), which is 6.3 kb, was derived by cloning the *cos* region into the nonessential region of plasmid

TABLE 1.5 Properties of Selected Cloning Vectors

Type	Name	Size (kb)	Selective Markers[1]	Cloning (Endonuclease) Site[2]	Host	Comments	References
I. PLASMIDS							
(a) General Purpose							
	pBR322	4.3	Ap[r],Tc[r]	AvaI,PstI,BamHI,PvuII, ClaI,SalI,EcoRI,HindIII, PvuI	E. coli	Complete sequence known. Certified EK2 vector. PvuI, PstI site in Ap[r]. BamHI, HindIII, SalI in Tc[r]. Amplified by chloramphenicol. See Figure 1.6.	Bolivar et al., 1977b; Sutcliffe, 1979.
	pACYC184	4.0	Cm[r],Tc[r]	BamHI,EcoRI,HindIII, SalI	E. coli	EcoRI site in Cm[r]. BamHI, SalI in Tc[r]. Chloramphenicol amplifiable.	Chang & Cohen, 1978.
	pMB9	5.8	Tc[r],ImmEI	BamHI,EcoRI,HindIII, SalI	E. coli	BamHI, HindIII, SalI in Tc[r]. Used mainly for EcoRI cloning. Chloramphenicol amplifiable. Certified EK2 vector.	Bolivar et al., 1977a.
	pUB110	4.5	Km[r]	BamHI,TacI,EcoRI, BglII,PvuII,XbaI	B. subtilis	BglII site in Km[r]. Certified HV2 vector.	Gryczan et al., 1978.
	pBD8	6.0	Km[r]Cm[r]Sm[r]	HindIII,BamHI,BglII, EcoRI,XbaI	B. subtilis	Recombinant of pUB110 and another HV2 vector (pSA2100). HindIII, EcoRI, sites in Sm[r], BglII site in Km[r]. Replication temperature sensitive.	Dubnau et al., 1980; Ehrlich et al., 1982.
	YIp5	5.4	Am[r]Tc[r],Ura	EcoRI,HindIII,BamHI, SalI	S. cerevisiae	BamHI, SalI sites in Tc[r]. Carries sufficient yeast chromosome DNA for integration via homologous recombination. Certified HV2 vector. See Chapter 2.	Hinnen et al., 1978; Struhl et al., 1979.
	YEp6	7.9	Am[r],His	EcoRI,XhaI,SalI	S. cerevisiae	XhoI site in His. Carries replication region of 2 μm yeast plasmid. Certified HV2 vector.	Broach et al., 1979; Struhl et al., 1979.
	YRp6	5.7	Am[r],Tc[r],Trp	BamHI,SalI	S. cerevisiae	BamHI, SalI in Tc[r]. Carries origin of DNA replication from yeast allowing autonomous replication of vector in host. Certified HV2 vector. See Chapter 2.	Kingsman et al., 1979; Struhl et al., 1979
	pKT263	12.8	Km[r],Sm[r]	HindIII,XmaI,XhoI, ClaI, EcoRI,SstI	P. aeruginosa	HindIII, XmaI, XhoI in Km[r], EcoRI, SstI in Sm[r]. Certified HVI vector.	Bagdasarian et al., 1981
	pRO1614	5.8	Tc[r]Cb[r]	PstI,BamHI,HindIII	P. aeruginosa	BamHI,HindIII sites in Tc[r].	Olsen et al., 1982
(b) Special Purpose							
	pKH47	4.4	Ap[r]Tc[r]	(Same as pBR322 less PvuII)	E. coli	All properties of pBR322 plus insertion of poly(dA) poly(dT) at PvuII site that permits separation of denatured, linear strands by chromatography on oligo(dA) or oligo(dT) columns for sequence analysis.	Hayashi, 1982
	pHA105	2.4	ImmEI	EcoRI	E. coli	Chloramphenicol amplifiable. All vector-specific polypeptides less than 16,000 MW, thus facilitating analysis of larger proteins coded by cloned DNA in minicells.	Avni and Markovitz, 1979
	pHSV-106	7.8	Am[r],TK	ClaI,PvuI,BglII,KpnI, HindIII,BglI,AvaI	E. coli, eukaryotes	All properties of pBR322, plus carries Herpes simplex virus (type 1) thymidine kinase gene inserted in BamHI site. Useful for expression and regulation studies in prokaryotes, eukaryotes. PvuII site in Am[r], KpnI in TK.	McKnight and Gavis, 1980
	pUC8/9	2.7	Am[r]Lac	EcoRI,AvaI,BamHI, SalI,HincII,AccI, PstI,HindIII,SmaI	E. coli	Carries multiple cloning sites of phage M13mp8/9 inserted into HaeII site of pBR322. All cloning sites in Lac. See Figure 1.9.	Vieira and Messing, 1982
II. COSMIDS							
	pHC79	6.4	Am[r]Tc[r]	EcoRI,ClaI,BamHI, SalI,PstI,PvuI	E. coli	Derivative of pBR322 carrying lambda cohesive site cos. Useful for cloning DNA up to 40 kb into gene banks. See Figure 1.7.	Hohn and Collins, 1980
	pYC1	24.4	Am[r]His	BamHI,EcoRI,XhoI, SalI	S. cerevisiae	Derivative of pHC79 and YEp6. XhoI site in His. Replicates autonomously in yeast.	Hohn and Hinnen, 1980
	pKT264	11.2	Am[r]Sm[r]	EcoRI,SstI	P. aeruginosa	Related to pKT263. Cloning sites in Sm[r]. Certified HVI.	Bagdasarian et al., 1981

(continued)

TABLE 1.5 *(Continued)*

Type	Name	Size (kb)	Selective Markers[1]	Cloning (Endonuclease) Site[2]	Host	Comments	References
III. BACTERIOPHAGE							
(a) Lambda (λ)							
	Charon 28	39.4		*Eco*RI,*Bam*HI,*Hind*III, *Sal*I,*Xho*I	*E. coli*	Useful for cloning DNA fragments up to 19 kb into gene libraries.	Rimm *et al.*, 1980
	λ1059	45	Grows in RecA⁺, P2 lysogens	*Bam*HI	*E. coli*	Eliminates need for purifying phage arms. Carries plasmid *Col*EI replication origin, thus can multiply nonlytically. Useful for cloning fragments from random *Sau*3A cuts.	Karn *et al.*, 1980
(b) M13							
	M13mp7	7.237	Lac	*Acc*I,*Bam*HI,*Eco*RI, *Hind*II,*Pst*I,*Sal*I	*E. coli*	Endonuclease sites in *lacZ* gene (multiple cloning sites). Useful for preparing single-stranded DNA for sequencing. Complete sequence known.	Messing *et al.*, 1981; Van Wegenbeck *et al.*, 1980
	M13mp8/9	7.229/ 7.599	Lac	(All mp7 sites plus *Hind*III,*Sma*I,*Xma*I)	*E. coli*	Cloning sites in opposite orientations in mp8, mp9 allowing insertion of cloned DNA in two orientations	Messing and Vieira, 1982

1. Am^r: resistance to ampicillin, Tc^r: tetracycline, Cm^r: chloramphenicol, Sm^r: streptomycin, Km^r: kanamycin, Cb^r: carbinicillin ImmEI: immunity to colicin EI Ura: uracil synthesis, Trp: tryptophan synthesis, His: histidine synthesis, TK: thymidine kinase, Lac: lactose fermentation
2. Only unique sites given

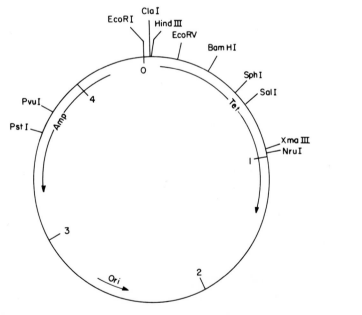

FIGURE 1.6 Plasmid cloning vector pBR322.

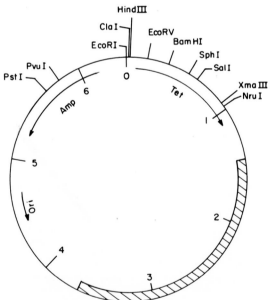

λ DNA

FIGURE 1.7 Cosmid pHC79.

pBR322 (Hohn and Collins, 1980). It retains all the unique endonuclease sites of pBR322 and insertional inactivation may be used to select recombinant molecules.

Although cosmids were designed to facilitate the cloning of large fragments of DNA by way of *in vitro* phage packaging, it is possible, by utilizing the appropriate λ-derivative, to achieve *in vivo* (within bacterial host) packaging as well. Thus, large but limited regions of DNA isolated in this manner can be subjected to genetic manipulations, such as transposon-mediated mutagenesis, that occur only *in vivo* (White *et al.*, 1983). Klee *et al.* (1983) used transposons Tn3 and Tn5 in this manner to analyze the virulent region of the tumor-inducing plasmid of *Agrobacterium tumefaciens*.

Two types of *E. coli* DNA phages, i.e., the temperate (e.g., λ) and filamentous (e.g., M13) phages, are used extensively as prokaryotic cloning vectors. Lambda phage contains a linear duplex DNA 50 kb in size. Numerous genetic manipulations by various laboratories have produced many (upwards to 100) λ derivatives useful as cloning vectors (for a review, see Williams and Blattner, 1980). The insertional-type vectors carry only one restriction endonuclease cleavage site that permits cloning by the addition of foreign DNA to form recombinant phages which are up to 105% of the phage's original size. Substitution vectors with two or more endonuclease sites allow for the replacement of up to 25% of the phage DNA by foreign DNA. Examples of the latter vector-type are the derivatives λgtWES·λB used for cloning *Eco*RI fragments (2−15 kb in length, Tiemeir *et al.*, 1976; Leder *et al.*, 1977), and Charon 28 which is used primarily for *Bam*HI cloning (6−19 kb fragments, Liu *et al.*, 1980) but which also carries *Hin*dIII, *Eco*RI, *Sal*I and *Xho*I sites (Rimm *et al.*, 1980). For effective cloning in λ, only the lytic genes at the ends of the genome must remain intact. Thus DNA cloning sites are found in the central regions of the genome. Recombinant phage can be selected and screened in numerous ways depending on the particular derivative used. Among these are size selection due to packaging requirements (e.g., λgtWES·λB), lactose or biotin metabolism (e.g., Charon 3A,4A,16A; Figure 1.8), resistance to chelating agents, plaque morphology, genetic complementation, restriction endonuclease analysis and nucleic acid hybridization (e.g., Charon 28; see Williams and Blattner, 1980, for a review). Lambda vectors are generally used for building gene libraries or DNA banks (Blattner *et al.*, 1978; Maniatis *et al.*, 1978). Because packaging restraints dictate a minimum DNA length for viable particle formation, the cloning of small DNA's (1 kb) in λ, which would account for only 1−2% of the DNA recovered, is much less efficient than cloning with small plasmids, such as pBR322, where cloned fragments would constitute 20% of the DNA. Derivatives λgtWES·λB and Charon 3A,4A,16A,21A,23A are certified EK2 vectors (see Maniatis *et al.*, 1982, for a recent listing of their properties).

The filamentous phage M13 carries a single strand of DNA 6407 nucleotides in length. Its small size relative to phage λ, and the ease with which high titres (50−100 copies per cell) are achieved, make M13 an attractive vector for cloning. Further, its single-strandedness allows for the direct nucleotide sequencing of the cloned DNA

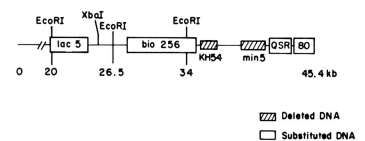

FIGURE 1.8 λ Cloning Vector Charon 4A.

by the dideoxy method (Sanger *et al.*, 1977). For in-depth considerations of the filamentous phages as cloning vectors see Barnes (1980) and Zinder and Boeke (1982). Messing and coworkers (1977, 1981, 1982) have developed a series of M13 derivatives specifically for cloning purposes. The derivative M13mp7 carries the promoter and the N-terminal portion of the *E. coli* β-galactosidase gene. The β-galactosidase gene, in turn, carries a DNA insert containing an array of restriction endonuclease sites (multiple cloning sites, MCS). Insertions of *Eco*RI, *Bam*HI, *Sal*I, *Acc*I, *Pst*I or *Hinc*II-generated fragments into the MCS results in the loss of β-galactosidase activity thereby permitting the direct selection of recombinant particles. Derivatives mp8 and mp9 both contain all the restriction sites of mp7 (reduced in number from two to one) plus new sites for *Hind*III, *Sma*I and *Xma*I but in opposite orientations. The ability to vary the orientation of a cloned fragment is of value in DNA sequence analysis (see Section XVI of this chapter for more detail), transcriptional mapping and *in vitro* mutagenesis studies. The M13mp phages have the disadvantage that certain inserts are unstable depending on the insertion site employed (Barnes, 1980). All M13mp derivatives are suitable for cloning in EK1 host-vector systems.

Vieira and Messing (1982) inserted the multiple cloning site segment from phages M13mp8 and mp9 into a slightly modified derivative of pBR322 to produce a set of plasmids, pUC8 and pUC9 (see Figure 1.9), that have proven very useful in the cloning of doubly-digested restriction fragments in opposite orientations.

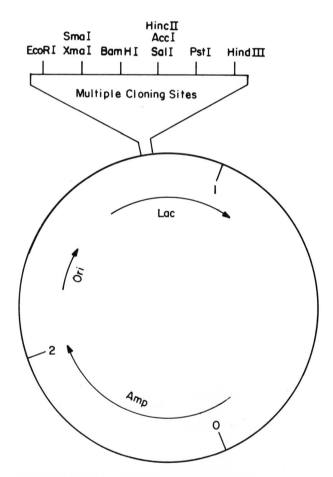

FIGURE 1.9 Plasmid vector pUC8 containing the multiple cloning site segment of bacteriophage M13mp8.

C. EXPRESSION VECTORS

The vectors used for the isolation and amplification of a particular fragment of DNA may not necessarily be the best for its expression. A number of alternative vectors have been designed specifically for this purpose. To serve as an expression vector, the molecule must carry, in addition to all the requirements of a good cloning vector, those DNA sequences essential for the transcription and translation of cloned genes. For successful expression, the coding region of the gene under study must be continuous, not dispersed. The gene in question must come under the control of a host promoter that is efficiently recognized by the host RNA polymerase. Promoters that are strong and at the same time, well-regulated, are most useful. Regulation is essential since continuous high-level expression may be detrimental to the host cell. The strong leftward promoter, *pL* of bacteriophage λ (e.g., plasmid pKC30, Shimatake and Rosenberg, 1981; Rosenberg *et al.*, 1983), the *lac* promoter of *E. coli* (e.g., plasmid p*β-gal*13C of Goeddel *et al.*, 1979), the *trp* promoter of *E. coli* and *Serratia marcesens* (e.g., plasmids p*trp*L1 of Edman *et al.*, 1981 and the pBN series of Nichols and Yanofsky, 1983) and the promoters of penicillin resistance genes (e.g., plasmid pKT287 of Talmadge *et al.*, 1980; plasmid pSYC193 of Chang *et al.*, 1982) are such promoters. The *pL* promoter is generally considered to be several fold more efficient than the *lac* promoter (McKenney *et al.*, 1981; Remault *et al.*, 1983).

As well, the mRNA product should be fairly stable and efficiently translated. The latter is ensured by the presence of ribosomal binding sites that include both an initiator codon and an upstream Shine-Dalgarno sequence (Shine and Dalgarno, 1975). Numerous attempts have been made at constructing vectors satisfying these requirements. In each instance, insertion of the gene(s) to be cloned was accomplished by fusing them with genes present on the vector. Thus, inserted genes have been fused with the *N*-terminal portions of a plasmid gene (the β-galactosidase gene in plasmid p*β-gal*13C, Goeddel *et al.*, 1979), with an *E. coli* ribosome-binding site (e.g., Roberts *et al.*, 1979), or with artificial ribosome-binding sites (Chang *et al.*, 1978). Bernard and Helinski (1980) have described a number of these in detail.

A series of promising expression vectors was created by introducing the *pL* region of λ into a derivative of a replication-control mutant of plasmid R1 (Remault *et al.*, 1983). Heat induces an increase in plasmid copy numbers thereby resulting in gene amplification. The *pL* promoter is controlled by a temperature sensitive repressor whose determinant, λ gene *cI*, is located on a second, compatible plasmid. Thus, exposure to elevated temperatures concomitantly turns on the λ promoter. (Activation of the *lac* and *trp* promoters is achieved through the addition of particular inducer molecules to culture media.)

D. SHUTTLE VECTORS

Shuttle vectors carry DNA sequences which allow them to replicate in different genera of prokaryotic organisms or in both prokaryotic and eukaryotic host cells. They can, therefore, be used to shuttle DNA back and forth between such cells as *Escherichia coli* and *Streptococcus* sp. (Macrina *et al.*, 1982) *E. coli* and *Saccharomyces* sp. (Larsen *et al.*, 1983), *E. coli* and *Streptomyces* (Harris *et al.*, 1983) and *E. coli* and mammalian cells (DiMais *et al.*, 1982; Calos *et al.*, 1983). Using such vectors a particular DNA may be introduced with relative ease into a genetically divergent organism where its fate (replication, degradation, modification, transcription, expression, etc.) may be studied in detail. Further, this DNA may be recovered from the "adoptive" cells and returned, via plasmid-mediated transformation, to the natural host where it may be amplified and its activity assayed by traditional methods.

The bacterial-mammalian shuttles are particularly valuable. Plasmids pSVi2 and pBPV-HII, which contain essential replication functions derived from plasmid pBR322, SV40 DNA (Calos et al., 1983) and bovine papilloma virus (DiMais et al., 1982), respectively, have been used as vehicles for the introduction and replication of specific (marker) DNA into mammalian cells. The pBR322 derivative, plasmid pML2 (Am^r, 2.9 kb) is used in constructing these E. coli-mammalian shuttles because it lacks the plasmid sequences that are inhibitory to virus replication in simian (Lusky and Botchan, 1981) and mouse (Sarver et al., 1982) cells. Marker DNA replicated in mammalian cells as part of such vectors is highly susceptible to mutagenesis. For instance, the galK (Razzaque et al., 1983) and lacI (Calos et al., 1983) genes placed in simian cells displayed mutation frequencies of $1-2 \times 10^{-2}$ as compared to spontaneous mutation rates of 10^{-6} to 10^{-7}. Thus, not only do shuttle vectors provide a means for introducing foreign DNA into eukaryotic cells, they appear to be potentially useful in studies of mutation at the DNA sequence level in these cells as well.

E. HOST ORGANISMS

Microorganisms commonly used as hosts for cloning experiments include the gram-negative rod, Escherichia coli, the gram-positive spore-forming rod, Bacillus subtilis and the yeast Saccharomyces cerevisiae. (For details on the latter two, see Dubneau et al., 1980 and chapter 3 of this manual, respectively.) Ideally, strains used for this purpose should be free of any restriction enzyme activity that might degrade foreign DNA. If such derivatives are not available, damage may be minimized by extracting DNA initially from modification-plus propagating organisms since methylation offers some protection against restriction enzymes. Further, it is often advantageous that the strain be defective in homology-based recombination (recA) since inadvertant recombinational events may threaten the integrity of the cloned DNA. (This option is also available for λ-based vectors through the red gene; see Leder et al., 1977.) Suitable hosts for bacteriophage vectors must also carry genes capable of supressing debilitating mutations, such as those affecting particle formation (W,E genes of λ) or host lysis (S gene of λ), that are "built-in" to such vectors for biosafety reasons (e.g., supF for λgtWES·λB hosts, supE for M13mp-hosts). Finally, the nucleic acid of the host should be easily separable from that of the cloning vehicle by physical means (e.g., differential centrifugation based on G-C content or encapsulation), or differ in its reactivity to alkali or organic acids (phenol). The E. coli strains C600 (F$^-$thi-1 thr-1 leuB6 lacY1 tonA21 SupE44, a.k.a. CR34, Appleyard, 1954), HB101 (F$^-$hsdS20[r$_B^-$m$_B^-$] recA13 ara-14 proA2 lacY1 galK2 λ$^-$ strA, Bolivar and Backman, 1979), and RR1 (HB101 rec$^+$, Bolivar et al., 1977a & b) are commonly used plasmid hosts that meet all or most of these criteria. E. coli strain LE392 (F$^-$hsdS20[r$_K^-$m$_K^-$] lacY1 galK2 metB1 trpR55 SupE44 SupF58 λ$^-$, Murray et al., 1977) and JM103 (F'[traD36] Δlacpro thi endA sbcB15 hsdR4 SupE strA, Messing et al., 1981), respectively, satisfy requirements as λ and M13 hosts.

Depending on the nature and origin of the DNA to be cloned, current biohazard guidelines (Fed. Regist. 48:24556, 1983) may require the use of host organisms severely limited in their survival capacities. Only a small number of strains have been certified for this purpose. E. coli strain χ1776 carries numerous chromosomal mutations (e.g., dapD8, thyA142, metC65, cycA1) and two deletions (gal-uvrB, bioH-asd) that render the cell so "sick" that growth is achieved only under highly specific laboratory conditions (Curtiss et al., 1977). It is also F$^-$ (no fertility plasmid) and does not permit λ or T-phage attachment. Because of these properties, χ1776 is considered a suitable EK2 host for plasmid cloning. The E. coli strain DP50supF (a.k.a. χ2098), is a certified EK2 host for λ cloning. It has the genotype F$^-$ dapD8, *lacY1*, Δ (*gal-uvrB47*), *tyrT58*, *supF58*, Δ (*thyA57*) (Blattner *et al.*, 1977). The possession of *dapD8* (requires diaminopimelic acid), Δ*thyA* (requires thymine) and the *gal-uvrB* deletion (requires biotin) ensures the low survival of this strain in nature. Both χ1776 and DF50*supF* are defective in their restriction-modification systems (*hsd*S/R). Thus only a very low number of the plasmids or phages propagated in them would survive in bacteria from nature that have a restriction modification system similar to *E. coli* K12.

If expression of the cloned DNA is to be measured, it may be convenient to have the hybrid molecule in a bacterium that produces chromosome-free mini-cells. The latter usually retain plasmid (cosmid) DNA and sufficient enzymes etc., to allow transcription and translation. Strain χ1776 is one such mini-cell producer.

Information regarding cloning in *Pseudomonas*, *Streptomyces*, *Neurospora*, *Bacillus* and *Saccharomyces* species and mammalian hosts may be obtained from Hofschneider and Goebel (1982), Ganesan *et al.* (1982) and chapters 3 and 4 of this manual. U.S. National Institutes of Health certified hosts are available for all the bacterial systems listed (Fed. Regist. 48:24556, 1983).

METHOD 1: CLONING USING RESTRICTION-ENZYME-GENERATED COHESIVE OR BLUNT ENDS

Cleavage of circular DNA with certain restriction endonucleases produces linear molecules with either 5' or 3' extensions at their ends. If two separate molecules are cut with identical enzymes, then mixed together in a manner which optimizes the chances of these termini annealing by hydrogen bonding, a recombinant molecule may be formed. The inserted molecule may be oriented in either of two directions. Specific orientation of the fragment may be achieved by using a technique called **directional cloning**. Digestion of a particular DNA molecule with two separate restriction endonucleases will generate DNA fragments having dissimilar termini. If the vector and the DNA to be cloned are digested with the same two enzymes and then mixed, the vector and foreign DNA fragments will anneal only in one direction. For example, the vector pBR322 and the DNA to be inserted are digested with both *Hin*dIII and *Cla*I: the separately generated *Cla*I (or *Hin*dIII)-termini will anneal with each other, but not the *Cla*I and *Hin*dIII termini, thereby giving directionality to the cloning.

PROCEDURE

1. Digest separately, 0.25–0.5 μg of vector DNA (e.g., pBR322) and 0.5–1μg of the DNA to be cloned with the "appropriate" restriction enzyme using conditions outlined in Method 1, Section IX, p. 36.

COMMENTS

1. The procedure described here is a **"shotgun"** approach. Alternatively, a specific DNA fragment can be isolated by gel electrophoresis and extracted before cloning into the vector. In such cases, less (about 0.1 μg) DNA is required. The appropriate restriction endonuclease is determined by the specific cloning vector used. Two different enzymes may be used to cleave the plasmid vector to minimize its tendency to spontaneously recircularize before foreign DNA is inserted. These same two enzymes must be used for cleaving the DNA fragment to be inserted (see Method 2, p. 52, of this section).

PROCEDURE	**COMMENTS**

In this regard, it is important to note that the two enzymes may give the fragments with identical termini. For example, *Taq*I cleaves the sequence

$$\downarrow$$
$$\text{T-C-G-A}$$
$$\text{A-G-C-T,}$$
$$\uparrow$$

while *Hpa*II recognizes the sequence

$$\downarrow$$
$$\text{C-C-G-G}$$
$$\text{G-G-C-C.}$$
$$\uparrow$$

Since the 5′ termini produced by these enzymes are identical, they may be annealed and ligated, producing

$$\text{T-C-G-G}$$
$$\text{A-G-C-C.}$$

However, the new sequence generated is not recognized by either enzyme.

2. Add 400 μL of 0.3M sodium acetate. Add an equal volume of TE-saturated phenol, mix by inversion and centrifuge to separate the phases. Remove the upper aqueous layer to a new 1.5 mL microfuge tube.

2. See this chapter, Section XVII-F, p. 98, for the preparation of saturated phenol solutions.

3. Add 500 μL ether, mix, centrifuge and remove the upper organic layer with a Pasteur pipette. Remove residual ether by evaporation under a stream of air.

3. Residual phenol must be removed with ether as it will denature enzymes used in subsequent steps.

4. Add two volumes of ice-cold ethanol and mix gently. Chill at −70°C for 15 minutes or −20°C for 2 hours. Pellet the DNA by centrifugation at full speed in a microfuge for 5−15 minutes. Remove the supernatant and dry the pellet either under vacuum or by inverting the tube on a rack (air dry).

5. Suspend the two DNA fragments in 10 μL of ligation buffer, then mix together (final volume = 20 μL).

5. Ligation Buffer

20 mM Tris-HCl, pH 7.6
10 mM MgCl$_2$
10 mM dithiothreitol
0.6 mM ATP

Dithiothreitol is a water-soluble reducing reagent which reduces disulfides to sulfhydryl (SH) thus protecting the ligase enzyme from oxidation.

6. Add 0.01−0.1 units of T4-DNA ligase. Incubate at 12°C for 2−18 hours.

6. There are two prokaryotic DNA ligases: a NAD-dependent enzyme from *E. coli* and an ATP-dependent enzyme coded by bacteriophage T4. Both enzymes join fragments with overlapping, complementary termini (sticky ends). However, only T4 ligase can join fragments with "blunt ends". Optimum conditions for ligating DNA fragments have been described by Dugaiczyk *et al.* (1975). Although the ligase has an optimum temperature of 37°C, A-T rich hydrogen-bonded termini, such as those produced by digestion with *Eco*RI have a melting temperature of 5–6°C. Thus, a compromise temperature of 10–15°C is generally chosen for ligation reactions. Other endonucleases (e.g., *Hin*dIII) generate stronger cohesive ends and therefore have a higher melting temperature. They may be ligated at higher temperatures. The ligation step can be monitored by subjecting small aliquots to agarose gel electrophoresis.

7. Transform an antibiotic sensitive *E. coli* K12 with the ligated mixture as outlined in Section XIV, p. 81.

7. Use the transformation procedure outlined in Section XIV, p. 81. Strains, such as EC490 (Met⁻ Lac⁻), HB101 (Pro⁻, Leu⁻, Lac⁻), and C600 (Thi⁻, Thr⁻, Leu⁻, Lac⁻), which lack restriction and modification systems will give better transformation frequencies.

8. Enrich for transformants by using the appropriate selection plates.

8. In this case, the cloning vector should carry more than one selectable phenotype (e.g., pBR322 is resistant to tetracycline and ampicillin). When DNA is specifically inserted into one of the markers, its expression will be interrupted. For example, a DNA insert in the tetracycline gene of pBR322 will "inactivate" the gene and can be screened by looking for ampicillin-resistant and tetracycline-sensitive transformants. Thus, the selection employed is determined by the site of insertion. For example, if the *Bam*HI site of pBR322 is used for insertion (i.e., tetracycline is inactivated), ampicillin plates (50 μg/mL) should be used. If the *Pst*I site is used (i.e., the ampicillin gene) select on tetracycline plates (25 μg/mL).

Other phenotypes may be screened by **DIRECT (POSITIVE) SELECTION**. For example, cloned DNA coding for lactose-fermentation can be directly screened on lactose-containing indicator plates, such as MacConkey plates, providing that the host strain is Lac⁻.

9. Pick individual colonies and check for markers originally coded by the vector.

9. In the case of insertional inactivation of pBR322, check for resistance to both tetracycline and ampicillin. If the transformants are resistant to only one of the two antibiotics, it is then quite safe to assume that foreign DNA is inserted into the plasmid vector.

10. Maintain the recombinant clones for further studies.

10. If very large fragments of DNA have been cloned, it may be desirable to **SUBCLONE** (i.e., reclone a smaller piece of the fragment which still possesses the gene of interest).

METHOD 2: MINIMIZING THE RECIRCULARIZATION OF PLASMID VECTORS WITH ALKALINE PHOSPHATASE

Two procedures are commonly used to minimize the recircularization of a restriction endonuclease-digested vector prior to the addition of foreign DNA. In the first approach, the vector is digested with two separate enzymes which produce different cohesive ends. *Eco*RI and *Hin*dIII or *Bam*HI and *Hin*dIII may be used for plasmid pBR322, for example. Because the cohesive ends comprise different base sequences, they cannot spontaneously reanneal to form circular molecules. The addition of the DNA fragment to be cloned, previously digested with the same two enzymes and therefore having the same cohesive ends, will result in spontaneous annealing.

The second method involves the use of alkaline phosphatase and is based on the premise that DNA ligases cannot function on molecules with the 5′ phosphate removed. Alkaline phosphatase is used to remove the 5′ phosphate from linearized (i.e., digested) vectors thereby preventing them from ligating. The DNA fragment to be inserted is not exposed to the enzyme. Because it has a normal 5′ phosphate group, it can be ligated to the ends of the plasmid vector, thereby forming an open circular recombinant molecule. This molecule may then be transformed into the appropriate host.

PROCEDURE

1. Digest 0.25 – 0.5 μg of the cloning vector with the appropriate restriction enzyme, using conditions outlined in Section IX, p. 36.

2. Dephosphorylate the linearized DNA with alkaline phosphatase by adding the following into a 1.5 mL microfuge tube:

a. Approximately 1 μg of the DNA fragment (volume 5 μL).

b. 25 μL of 50 mM Tris-HCl (pH 8.0) buffer.

c. 20 μL(2 – 5 unit/mL) alkaline phosphatase.

d. Incubate at 37°C for 30 – 60 minutes.

3. Add 400 μL of 0.3M sodium acetate. Add an equal volume of saturated phenol, mix by inversion and centrifuge to separate the phases. Remove the upper aqueous layer to a new 1.5 mL microfuge tube and add 500 μL ether. Mix, centrifuge and remove the upper organic layer containing any residual phenol.

4. Add two volumes of ice-cold ethanol and mix gently. Chill at −70°C for 15 minutes or −20°C for 2 hours. Pellet the DNA by centrifugation at full speed in a microfuge for 5 – 15 minutes. Remove the supernatant

COMMENTS

1. It may be helpful to check that digestion is complete by subjecting a small aliquot (10 μL) of the reaction mixture to agarose gel electrophoresis.

c. Either bacterial alkaline phosphatase or calf intestinal alkaline phosphatase may be used. The latter has about 10 – 20 fold higher activity per mg of protein than the bacterial enzyme (Chaconas and Van de Sande, 1980).

3. The phenol extraction removes alkaline phosphatase. Residual ether can be removed by passing a stream of air over the surface of the aqueous layer.

and dry the pellet either under vacuum or by inverting the tube on a rack (air dry).

5. Digest 0.25−0.5 μg of the DNA to be cloned using the same enzyme as in Step 1.

5. This step can be performed simultaneously with step 1.

6. Phenol and ether extract, then precipitate the DNA as described in steps 3 and 4.

7. Resuspend the dephosphorylated vector and the DNA to be cloned in 10 μL of ligation buffer, and mix. (Final volume = 20 μL).

7. See Method 1, p. 50, of this section for ligation buffer.

8. Add 0.01−0.1 units of T4-DNA ligase to the mixture and incubate at 12°C for 2−18 hours.

9. Transform competent *E. coli* cells.

9. Use the transformation procedure outlined in Section XIV, p. 81.

METHOD 3:   HOMOPOLYMERIC TAILING

The enzyme terminal transferase (terminal nucleotidyl transferase, Bollum, 1962; Chang and Bollum, 1971) acts by sequentially adding deoxynucleotide residues to the 3′ ends of linear DNA molecules without using a template in the presence of cacodylate buffer (Kornberg, 1974). Blunt-ended DNA molecules are inefficient substrates for this enzyme (Lobban and Kaiser, 1973), however, the substitution of magnesium with cobalt ion in the buffer permits the extension of duplex termini (Brutlag *et al.*, 1977). Because the enzyme is nonspecific, any of the four dNTP can be used as precursors thereby generating 3′ tails of a single type of residue (homopolymeric tail). For example, poly(dG) tails can be built at the 3′ ends of a DNA fragment (i.e., a cloning vector), while poly(dC) tails are generated at the 3′ ends of another DNA molecule (i.e., DNA fragment to be cloned). Since these two homopolymeric tails are complementary to each other, the two DNA fragments can be joined together by base pairing, and then joined together with ligase. Using this technique, Jackson *et al.* (1972) successfully cloned λ DNA into SV40. Subsequently, other genes such as proinsulin (Villa-Komaroff *et al.*, 1978), immunoglobulin gene (Seidman *et al.*, 1978), human globin gene (Wilson *et al.*, 1978) and yeast nuclear DNA (Petes *et al.*, 1978), were successfully cloned in the same manner.

DNA fragments cloned into a vector using homopolymeric tailing may be excised and recovered using one of the following strategies. First, the DNA fragment is inserted into the vector at a site which is closely flanked by two other restriction sites. For example, in pBR322, the *Cla*I site is located within thirty base pairs between *Eco*RI and *Hin*dIII sites. If a DNA fragment is inserted at the *Cla*I site with either poly(dA·dT) or poly(dG·dC) tailings, it can be excised and recovered by digesting the recombinant DNA molecules with both *Eco*RI and *Hin*dIII. The second strategy involves the regeneration of the restriction site at which the fragment is inserted by using the appropriate dNTP for tailing. For example, the *Pst*I site may be regenerated by poly(dG) tailings (Villa-Komaroff *et al.*, 1978) and theoretically the *Hin*dIII site can be regenerated by poly(dT) tailing (Bernard and Helinski, 1980). Thus in both cases, by digesting with the same enzyme (i.e., *Pst*I or *Hin*dIII), the cloned fragment can be excised and recovered. Figure 1.10 depicts how the *Hin*dIII site may be regenerated after poly(dT) tailing.

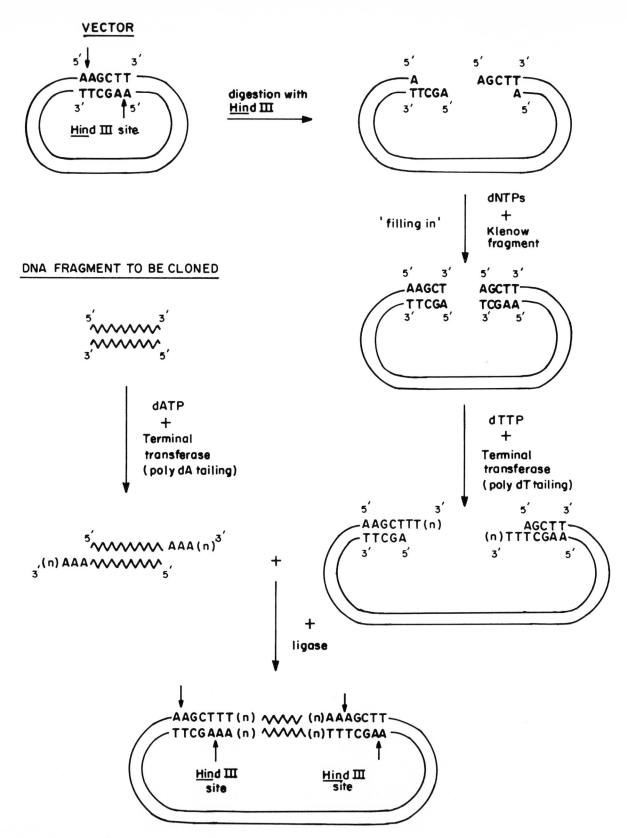

FIGURE 1.10 Regeneration of the *Hin*dIII restriction site by homopolymeric tailing. See Section XIIB, p. 70, for filling-in reaction.

PROCEDURE	COMMENTS
1. Completely digest 10 μg of vector DNA (e.g., pBR322) with an enzyme (*Pst*I) that gives a 3' overhang.	**1.** *Pst*I recognizes the following sequence:

↓

5'-C-T-G-C-A-G-3'
3'-G-A-C-G-T-C-5'

↑

and will give a 3' protruding end of:

5'-C-T-G-C-A-3'
3'-G

By adding homopolymeric (dG) tail to the 3' end, the recognition site of *Pst*I can be restored.

5' CTGCAGGGGG 3'
3' G

Duplex DNA molecules are always less efficient primers than single-stranded DNA. Duplex DNA molecules with 5' protruding ends are less efficient than those with 3' overhangs (Roychaudhury and Wu, 1980).

Partial digestion will result in a high background of nonrecombinant transformants after transformation of competent *E. coli* cells. Completely digested linear vector can be separated from undigested plasmid vector by agarose gel electrophoresis, then recovered by electroelution (Method 1, Section VIII, p. 31).

2a. Add 3M sodium acetate to a final concentration of 0.3M and precipitate the DNA with ethanol as described in the previous method. Resuspend the digested vector in 200 μL of tailing buffer containing 2 mM [α^{32}P]dGTP.

b. At the same time, ethanol precipitate and resuspend the DNA fragment to be cloned in 200 μL tailing buffer containing 2 mM [A^{32}P]dCTP.

2. 10× Tailing Buffer

1M cacodylic acid, sodium salt
250mM Tris-base (adjust to pH 7.6 with KOH)
  2 mM dithiothreitol
10 mM cobalt chloride

(To avoid precipitation, 0.1M cobalt chloride is added dropwise to the buffer with constant stirring. Roychaudhury and Wu (1980) observed that Co^{++} is more efficient than Mg^{+} for extending termini.)

3. Add 30–60 units of terminal transferase to the above mixtures. Remove 10 μL from each mixture and incubate at 37°C to determine the time required to incorporate 10–20 (dG) or (dC) residues. Keep the remainder of the samples at 0°C.

3. A sample of the reaction mixture is removed to determine the appropriate incubation time for tailing.

4. At 0, 1, 2, 3, 4 and 5 minutes, remove 1 μL aliquots from the sample incubated at 37°C and moni-

4. The number of dNTP residues added per 3' end may be calculated as follows:

PROCEDURE

tor for acid-insoluble radioactivity (Section XVII-O, p. 106). From the amount of radioisotope incorporated, determine the number of (dG) or (dC) residues which have been added to the ends of the vector and the DNA fragment to be cloned.

5. Incubate the remainder of the reaction mixtures at 37°C for the period of time that will give 10–20 (dG) or (dC) residues per end.

6. Chill to 0°C and add 20 μL of 0.5M EDTA or EGTA (pH 8.0).

7. Add an equal volume of TE-saturated phenol, mix by inversion and centrifuge to separate the phases. Remove the upper aqueous layer to a new 1.5 mL microfuge tube and add 500 μL ether. Mix and centrifuge. Remove the upper organic layer containing any residual phenol.

8. Separate the tailed DNA from unincorporated dNTPs by running the reaction mixture through a Sephadex G-50 or G-100 column (Section XVII-N, p. 103).

9. Add 3M sodium acetate to a final concentration of 0.3M then add two volumes of ice-cold ethanol and mix gently. Chill at −70°C for 15 minutes or −20°C for 2 hours. Pellet the DNA by centrifuging at 15,000 × *g* for 15 minutes. Remove the supernatant and dry the

COMMENTS

$$\frac{\text{counts per minute (cpm) per 1 } \mu L \text{ of acid-insoluble material (1)}}{\frac{\text{cpm per dNTP molecule (2)}}{} \times \frac{\text{total number of 3' ends per 1 } \mu L \text{ reaction mixture (3)}}{}}$$

(1) The cpm per 1 μL of acid-insoluble material is determined by the cold TCA method (Section XVII-O, p. 106).

(2) The cpm per dNTP molecule

$$\frac{\text{cpm of 1 } \mu L \text{ of } [\alpha P^{32}] \text{ dNTP stock solution}}{\text{concentration of dNTP (mole/}\mu L) \quad 6.023 \times 10^{23} \text{ (Avogadro's number)}}$$

(3) The total number of 3′ ends per (1) μL of reaction mixture (there are 2×3′ ends per DNA molecule)

$$= \frac{\text{amount of DNA (}\mu g) \text{ added to reaction} \times 6.023 \times 10^{23} \times 2}{\text{gram-molecular wt. of the DNA molecule} \times 10^6 \times \text{total volume of reaction mixture}}$$

6. Chilling reduces the activity of terminal transferase. EDTA or EGTA chelates the Co^{++} ion required for the reaction.

7. Phenol extraction removes terminal transferase.

8. The fractions are monitored either by scintillation counting (see Section XVII-O, p. 106) or with a Geiger counter. This step also eliminates small DNA fragments formed during the reaction which may interfere with cloning procedures.

PROCEDURE	COMMENTS

pellet either under vacuum or by inverting the tube on a rack (air dry).

10. Resuspend the tailed DNA in annealing buffer (~ 10 μl). Mix equimolar amounts of (dG)-tailed vector with (dC)-tailed DNA. (Final concentration 1 μg/mL; final volume ≃ 20 μL.)

10. Annealing Buffer

  10 mM Tris-HCl (pH 7.8)
 100 mM NaCl
    1 mM EDTA

11. Heat to 65°C for 5 minutes, then allow the DNA's to reanneal by incubating the mixture at 57°C for 1−2 hours.

12. Transform competent *E. coli* cells (Section XIV, p. 81).

METHOD 4: CLONING WITH SYNTHETIC LINKERS

Synthetic molecular linkers are oligonucleotides, 8−12 base pairs in length, having defined sequences that contain the recognition-cleavage site(s) of one or more restriction endonucleases, (e.g., *Eco*RI, *Bam*HI, *Hin*dIII, etc.). These linkers provide a convenient method for joining a blunt-ended DNA fragment (such as cDNA, see Section XV) to a cloning vector at a specific restriction site (Bahl *et al.*, 1976). If the DNA fragment to be cloned has sticky ends, it is transformed to blunt-ends either by filling-in or by 3′ → 5′ digestion mediated by the Klenow fragment of DNA polymerase I or T4 polymerase*. S1 nuclease, which digests single-stranded DNA, is not generally used to generate blunt ends for cloning purposes, because any nicked DNA would be digested to small fragments. The linkers are ligated to the blunt-ends of the DNA fragments with T4 ligase. The molecules are then digested with the appropriate restriction enzyme so as to produce the sticky ends, which can then be ligated to vectors with complementary ends. Although they may be chemically synthesized in the laboratory, this technology is generally beyond the capacity of most laboratories, and linkers are generally obtained from commercial sources.

PROCEDURE	COMMENTS

1. Digest 1 μg of the DNA to be cloned with the appropriate restriction endonuclease as outlined in Section IX, p. 35.

1. The appropriate enzyme is determined by the restriction site carried by a particular DNA. If the DNA molecule is already blunt-ended (i.e., after digestion with certain endonucleases, or if cDNA has been prepared), proceed directly to step 8.

2. Add 3.0M sodium acetate to a final concentration of 0.3M and precipitate the digested DNA with ethanol as described in previous methods.

3. Resuspend the DNA in 18 μL of polymerase buffer. If the DNA fragment has 5′ protruding ends, add 1 μL of a solution containing 2 mM of all four dNTPs. If the DNA has 3′ protruding ends, do not add the dNTPs.

3. With 5′ protruding ends, the "filling-in" reaction is carried out to produce blunt ends. (See Section XIIB, Method 1, p. 71.)

*(For a description of Klenow fragment activities, see Section XIIB, p. 70.)

PROCEDURE	COMMENTS

COMMENTS

Polymerase Buffer

   50 mM potassium phosphate (pH 7.4)
 0.25 mM dithiothreitol
   10 mM $MgCl_2$

PROCEDURE

4. Add 2.5 units of the Klenow fragment of *E. coli* polymerase I.

4. Like T4 polymerase, the Klenow fragment has $5' \rightarrow 3'$ polymerase and $3' \rightarrow 5'$ exonuclease activities. In the presence of all four dNTPs, the Klenow fragment will carry out its $5' \rightarrow 3'$ polymerase activity. In the absence of dNTPs, it will degrade the 3'- protruding ends of the DNA fragment (see Section XIIB, p. 70) with its $3' \rightarrow 5'$ exonuclease activity.

5. Incubate 10–15 minutes at 37°C.

6. Add 400 µL of 0.3M sodium acetate. Add an equal volume of saturated phenol, mix by inversion, and centrifuge to separate the phases. Remove the upper aqueous layer to a new 1.5 mL microfuge tube and add 500 µL ether. Mix and centrifuge. Remove the upper organic layer containing any residual phenol.

7. Add 2 volumes of ice-cold ethanol and mix gently. Chill at −70°C for 15 minutes or −20°C for 2 hours. Pellet the DNA by centrifugation at full speed in a microfuge for 5–15 minutes. Remove the supernatant and dry the pellet, either under vacuum or by inverting the tube on a rack (air dry).

8. Resuspend the DNA in 20 µL of ligation buffer.

8. Ligation Buffer

   20 mM Tris-HCl (pH 7.6)
   10 mM $MgCl_2$
   10 mM dithiothreitol
 0.60 mM ATP

9. Add 1–2 µg of synthetic linker and 1 unit of T4-DNA ligase.

9. Some manufacturers provide linkers with 5'-hydroxyl ends. Since T4-DNA ligase joins ends with 5'-phosphate group, the linker must be phosphorylated at the 5'-end using the forward reaction (Section XIIB Method 3, p. 75), before it is ligated to the DNA fragment.

10. Incubate at 12°C for 2 to 18 hours.

10. Longer incubation times (i.e., overnight) ensure better ligation.

11. Repeat steps 6 and 7.

12. Resuspend the DNA in restriction buffer and digest with a restriction enzyme, producing sticky ends.

12. The enzyme used depends on the specific recognition sequences on the linker molecule. High enzyme concentrations are required to ensure that digestion is complete.

ADAPTORS: The presence of identical restriction sites on the fragment to be cloned and the synthetic linker poses problems following digestion with that enzyme (i.e., the cloned fragment does not remain intact). This problem may be overcome by using synthetic adaptors which do not require digestion prior to cloning (Bahl and Wu, 1978). Adaptors are small oligonucleotides that have blunt-ends (i.e., double-stranded) at one end of the molecule and a single-stranded cohesive end at the other. The cohesive end carries a restriction endonuclease site. The adaptor molecule, which is phosphorylated at the 5′ blunt-end, but not the 5′ cohesive end, is ligated with T4 ligase to the blunt-ended DNA fragment to be cloned. The 5′ termini of the cohesive ends are phosphorylated prior to ligation to a vector with complementary termini.

13. Load the DNA onto a 1% agarose slab gel and electrophorese in order to separate restriction fragments from unincorporated linkers.

13. Unincorporated linkers will give a high background of nonrecombinant plasmids. Prepare and run the agarose gel as described in Section V, p. 18. This step may also be accomplished using a Sephadex G-100 or G-150 column (Section XVII-N, p. 103).

14. Visualize the DNA fragment either by ethidium bromide staining or by autoradiography, if the sample has been labeled with ^{32}P. Electroelute the fragment from the gel onto DEAE paper and recover the DNA as outlined in Section VIII, p. 31.

14. If the DNA molecule has been stained with ethidium bromide, it is necessary to remove the dye (Section VII, p. 29) before proceeding with restriction endonuclease digestions.

15. Digest the vector with the same restriction endonuclease.

16. Repeat steps 6 and 7.

17. Ligate the linker-modified (foreign) DNA molecule to the plasmid vector as outlined in steps 8–11.

18. Transform the appropriate *E. coli* cells (Section XIV, p. 81).

XI. Nucleic Acid Hybridization: Determination of Genetic Homology

When mixed, single-stranded DNA or RNA carrying specific nucleotide sequences in common, will anneal at regions of genetic homology to form duplexes. If the nucleic acids are from different sources, the duplexes are termed hybrid molecules. Thus, DNA/DNA hybridization may be used to test the relatedness of particular organisms. Further, the ability to form either DNA/DNA or DNA/RNA hybrid molecules can be exploited to locate particular genes in a given fragment of nucleic acid. Such hybridizations, used in combination with rapid DNA isolation (Section I, p. 1) and gel electrophoresis (Section V, p. 18) techniques, form the basis of several powerful methods for DNA analysis currently used in molecular biology.

For many of the hybridization reactions, DNA or RNA (that has been electrophoretically separated on a slab gel) is transferred to a membrane filter via capillary or electrical (electroblot) action prior to hybridization. The molecules are immobilized in the membrane matrix in exact replicas of their gel separations. Membrane papers (e.g., nitrocellulose, diazobenzyloxymethyl-DBM, DEAE) are preferred for the hybridizations, because they are more easily manipulated and stored than the fragile gels. Generally, the choice of paper used for transfer and hybridization is determined by such parameters as the specific nucleic acid, the size of the molecules, the number of steps involved in the process, and sensitivity. Nitrocellulose papers were initially used for transfers. However, the use of this paper was limited because it does not retain small (< 150 bp) DNA fragments, nor does it bind RNA. Recently, Smith and Summers (1980) found that the improved retention of small DNA fragments could be achieved by replacing SSC buffer with 1M ammonium acetate and 0.02N sodium hydroxide. Thomas (1983) has recently reported that RNA is easily transferred to nitrocellulose paper providing it is extensively denatured. Alwine *et al.*, (1977, 1979) have used diazobenzyloxymethyl-paper (DBM) for transferring and covalently binding DNA and RNA. Levy *et al.* (1980) found that small fragments of DNA are transferred efficiently to this paper. In this case, small DNAs are first separated on low concentration (< 4%) polyacrylamide gels (Frei *et al.*, 1983) prior to transfer by either capillary or electrophoretic forces. A further comparison of the different blotting papers is given in Table 1.6. Since Southern (1975) first described the transfer of DNA from agarose gels to nitrocellulose (Southern Blot), the colloquial name Northern Blot has been used in reference to the transfer of RNA to paper matrices while Western Blot (Burnette, 1981) refers to the electrophoretic transfer of proteins to such papers.

METHOD 1: SOUTHERN BLOTS

In the Southern procedure (Southern, 1975), the transfer of denatured DNA from agarose gels to nitrocellulose paper is achieved by capillary action. A buffer, sodium saline citrate (SSC), in which nucleic acids are highly soluble, is drawn up through the gel into the nitrocellulose membrane, taking with it single-stranded DNA which becomes "trapped" in the membrane matrix. This DNA is hybridized with a nucleic acid probe, and depending on the type of label, the hybridization bands of interest may be detected either by autoradiographic, fluorographic, or immunological methods.

In a variation of the capillary blotting procedure, it is possible to spot $1-5$ μL samples of test nucleic acid onto nitrocellulose paper pretreated with $20\times$ SSC (Kafatos *et al.*, 1979). The nucleic acid is fixed to the membrane by baking at 80°C for two hours, followed by prehybridization, hybridization and washing steps, which are essentially the same as those described for the Southern Blots (see below). Such **DOT BLOTS** can be used for immobilizing both small DNA's (< 200 nucleotides) and RNA. In the latter instance, fifty percent formamide must be included in the reaction solutions, and incubations must be completed at 42°C and 50°C for the hybridizations and washing, respectively (Thomas, 1980).

TABLE 1.6 Comparison of Blotting Papers

Paper	Advantages	Disadvantages
Nitrocellulose	binds ssDNA,RNA,protein capacity $80-100$ μg/cm^2 inexpensive minimal preparation	brittle, shrinks reuse limited because DNA not covalently bonded binding diminished at $10\times$ SSC special procedures to bind RNA or small DNAs

DBM/DPT	binds ssDNA,RNA,protein
	capacity $20-40$ μg/cm^2
	covalent immobilization of molecules
	successive assays possible using
	   different probes
		hybridization less efficient
		   than with nitrocellulose
		requires chemical activation
		time, temperature, pH labile
		expensive

DEAE	binds dsDNA,RNA capacity 15 μg/cm^2 quantitative recovery of DNA	limited capacity and versatility brittle
Nylon	binds DNA,RNA,proteins high detection sensitivity flexible heat and solvent resistant prewetting unnecessary	certain types may give high backgrounds

Because a large number of different nucleic acids may be spotted and immobilized on the same membrane for testing by a single probe, the dot blot procedure lends itself to rapid screening tests, and is particularly useful in the clinical laboratory. For example, it has been successfully employed as a sensitive and reproducible assay for the presence of rotavirus in human stool suspensions (Flores *et al.*, 1983). A modification of the procedure has also been used to detect and quantitate human cytomegalovirus in urine (Chou and Merigan, 1983).

PROCEDURE

1. Denature the DNA and transfer it to nitrocellulose paper as follows:

a. Immerse the gel in 500 mL of 0.4N NaOH and 0.8N NaCl (use 40 mL of 5N NaOH, 80 mL of 5N NaCl and 380 mL dH₂O to prepare the solution). Agitate gently for 30 minutes at room temperature. This and subsequent steps may be carried out in a Pyrex® baking dish.

b. Discard the denaturing solution.

c. Neutralize the gel by immersing it in 500 mL of 0.5M Tris-HCl and 1.5M NaCl (prepare this solution by mixing 250 mL 1M Tris-HCl, pH 7.6, 150 mL 5N NaCl, 100 mL dH₂O). Agitate gently for at least 30 minutes at room temperature.

d. Cut a piece of nitrocellulose paper such that its overall dimensions are 1 cm larger than the gel.

e. Using forceps, wet the nitrocellulose paper by completely immersing it in 500 mL dH₂O.

f. Add 500 mL of 20× SSC (final concentration 10× SSC). Soak for at least 15 minutes.

g. Wet a piece of (20 × 28 cm) Whatman 3MM filter paper in 10× SSC (use same solution as in step f). Then lay it over a 20 × 20 × 0.9 cm piece of

COMMENTS

a. This step denatures DNA converting it to a single-stranded form, thereby facilitating its binding to nitrocellulose paper and to subsequent hybridization. Large DNAs (> 10 kb) transfer less efficiently than small DNAs. To improve transfer, large DNA should be fragmented prior to denaturation either by acid depurination and base cleavage in 0.25N HCl for 15 minutes (Wahl *et al.*, 1979), or by shortwave uv irradiation of the DNA.

d. **Vinyl gloves must be worn throughout the entire procedure.** Oils and sweat on the skin's surface will "stain" the membrane, reduce transfer efficiency and cause absorption of the probe (Southern, 1979).

e. Certain preparations of nitrocellulose contain cellulose acetate which tends to weaken DNA binding to the membrane matrix. Therefore, the use of pure nitrocellulose is highly recommended (usually of 0.45 μ porosity).

Sometimes nitrocellulose paper will not absorb a high concentration solution such as 10× SSC. Thus, it is better to wet the nitrocellulose paper with water and then adjust to 10× SSC (final concentration).

f. **SSC Buffer (20×)**

3M sodium chloride
0.3M sodium citrate

20× SSPE (recipe in step 2b) may be substituted for SSC. For DBM or DPT paper a low pH buffer (1M NaOAC, pH 4.0) must be used in order to maintain the stability of the diazonium groups.

g. Air bubbles will interfere with DNA transfer by creating discontinuities in the paper wick. See Figure 1.11 for a drawing of the apparatus.

Plexiglas, supported by a 150 × 15 mm Petri dish, which in turn is set in a 30 × 37 cm Plexiglas buffer chamber. Make certain that no air bubbles are trapped beneath the filter paper. The ends of the filter paper should touch the bottom of the buffer chamber.

h. Center the denatured gel on the Whatman paper ensuring that no air is trapped underneath it.

h. A piece of plastic window screening may be used to support the gel while it is carried from one container to another.

i. Frame the gel with Plexiglas spacers of the same thickness and length as the gel. Leave 0.5 cm of space between the edges of the gel and the spacers.

j. Place the prewet nitrocellulose paper on top of the gel. Make certain no air is trapped between the membrane and the gel and that the membrane does not touch the 3MM paper beneath the gel. (The position of the spacers may be adjusted to prevent this). Remove air bubbles by gently rubbing the nitrocellulose layer.

j. Air trapped under the filter prevents the transfer of DNA from the gel to the membrane matrix.

k. Cover the nitrocellulose paper with two pieces of Whatman No. 1 filter paper (of same dimensions as the gel).

k. These papers ensure the even absorption of the elutant up through the gel.

l. Place a stack (about 13 cm) of absorbent papers over the filter paper and compress the absorbent paper by placing a Plexiglas plate (15 × 15 cm) on top of this. Weigh down with a heavy object (3–5 lbs).

l. Any highly absorbent paper will serve (e.g., a single package (75 sheets) of 15 × 15 cm paper table napkins).

One half of a clay brick provides an ideal weight.

m. Add sufficient 10× SSC (approximately 1L) to the reservoir to cover a good portion of the 3 MM filter paper wicks.

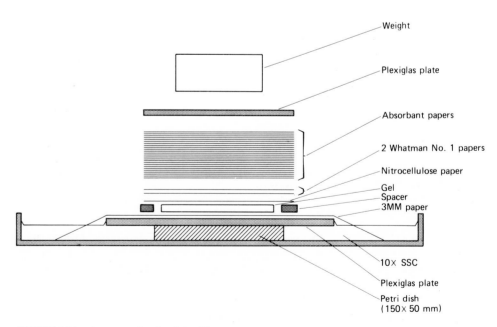

FIGURE 1.11 Apparatus for Southern Blots.

n. Make certain that the nitrocellulose paper does not touch the underlying 3MM paper in the gap between spacer and gel. If it does, bend the nitrocellulose upward using a spatula.

n. If the nitrocellulose touches the 3MM paper, the elutant buffer will pass directly from one filter to another rather than through the gel.

o. Leave overnight.

o. Depending on the filter paper used, several changes of absorbent material may be necessary. The length of time required for adequate transfer will vary with gel concentration, gel thickness, and DNA fragment size.

p. Remove the weight, absorbent material and Plexiglas spacers.

q. Turn the Whatman No. 1 filter paper and the nitrocellulose paper over, so that the gel remains attached to the nitrocellulose (this usually happens anyway). Place them gel side up on a piece of dry filter paper.

r. With a soft lead pencil trace an outline of the gel onto the nitrocellulose paper. Number the two outermost slots. Remove the gel.

r. This will facilitate the orientation of the autoradiograph relative to the original gel, thereby enabling identification of particular bands. The gel may be restained with ethidium bromide (Section VI) to determine whether the DNA has been transferred to the nitrocellulose.

s. Separate the nitrocellulose from the Whatman No. 1 paper and place it in 2× SSC. Rub gently to remove agarose debris. Pat dry between two filter paper sheets.

s. Residual agarose will interfere with subsequent hybridization reactions by giving fuzzy backgrounds.

t. Fix the DNA to the membrane by baking the nitrocellulose paper for 2–4 hours at 80°C under vacuum.

t. Because nitrocellulose paper tends to curl at the edges during this operation, it should be placed in the crease of a 3MM filter paper envelope which is fastened at the sides with cellotape. Baked filters may be stored at 4°C for several months.

2. Hybridize with a radioactive probe (see Section XII, p. 67) as follows:

2. Both DNA or RNA probes may be used. Since RNA does not bind to nitrocellulose, the prehybridization step (2b) may be omitted.

a. Place the nitrocellulose paper in a plastic bag (e.g. Sears Seal-N-Save bag) 3 cm larger than the paper. Heat seal one end with a commercially available electric sealer.

a. Bags may be purchased as open-ended continuous rolls (20 × 609.6 cm) thereby permitting variation in the length of the gels.

b. Place 10 mL of prehybridization solution (100 µg/mL single-stranded DNA and 1% sarkosyl in 5× SSPE) in the bag. Thoroughly wet the nitrocellullose paper by rubbing from the sealed to the open end. (This also serves to remove air bubbles.) Heat seal the plastic bag.

b. Use a plastic pipette to place the DNA solution in the plastic bag so that DNA will not stick to the pipette.

The prehybridization step is designed to block all the sites on the nitrocellulose paper that would bind the probe nonspecifically, thereby reducing the background.

20× SSPE

20 mM Na_2EDTA
0.2M NaH_2PO_4
3.6M NaCl
pH 7.0 (Adjust with 50% NaOH)

PREPARATION OF SINGLE-STRANDED DNA:
Dissolve 10 mg of heterologous DNA, such as herring sperm or calf thymus DNA, in 10 mL of water by stirring on a magnetic stirrer for 2 to 4 hours. Boil the solution for 10 minutes in a water bath to produce single-stranded DNA. To prevent reannealing, immediately cool the DNA solution by placing it on ice. The resultant single-stranded DNA can be added, in a ratio of 1:10, to 5× SSPE containing 1% sarkosyl (or SDS). The stock solution is stored at −20°C and boiled before use. Because single-stranded DNA tends to adhere to glass, a plastic bottle is used for denaturation and storage.

Since sarkosyl present in prehybridization solutions inhibits the binding of single-stranded DNA to nitrocellulose paper, a 5× solution of **Denhardt's Reagent** (Denhardt, 1966) may be used in its place.

Denhardt's Reagent (×50):

1.0% ficoll
1.0% polyvinylprollidone
1.0% bovine serum albumin
distilled H_2O
Filter, and store at −20°C.

c. Incubate in a 70°C water bath for 2−4 hours with gentle agitation.

c. Do not incubate in hot air chambers as they will dry out the membrane and crack it.

d. Boil the probe for 5−10 minutes and then immediately cool in an ice bath.

d. This step renders the probe single-stranded and may be initiated during the last 15 minutes of the prehybridization step. Probes prepared by nick translation (Section XII, p. 68) are suspended in TE buffer (pH 7.5).

e. Recover the membrane from the water bath, cut off one corner and pour the prehybridization solution into a plastic bottle. Force out any remaining solution by rubbing the plastic bag with a flat-edged object and collecting the solution on absorbent paper.

e. The prehybridization solution may be retained and used for subsequent hybridizations.

f. Add 5 mL of single-stranded radioactive probe and reseal the bag. Wet the nitrocellulose paper as described in step 2b to make certain that there are no dry spots.

f. Dry spots on the filter or the addition of probe solution to dry membranes may result in high backgrounds. With low concentration probes it may be advisable to add carrier (unlabeled heterologous) DNA to the solution (100 μg/mL). Do not use herring sperm or calf thymus DNA with probes homologous to higher animal DNAs. Alternatively, 10% dextran sulfate (MW 500,000) may be used in the presence of 50% formamide as a hybridization enhancer (reduces solubility of DNA) if the probe concentration is < 10 ng/mL (Wahl *et al.*, 1979).

g. Return the bag to the 70°C waterbath and incubate overnight with agitation.

g. Nucleic acids hybridize most efficiently at approximately 25°C below the temperature at which they are 50% denatured [melting temperature, (Tm) Studier, 1969]. Thus, for DNA in aqueous solutions ($2\times$ SSC), hybridizations are conducted at 65−70°C. Subsequent washings are completed at the same temperatures. Since formamide lowers the Tm (Hutton, 1977), hybridizations involving this reagent (50% solution) are conducted at 37−42°C, with washings at 50°C. Formamide also slows the rate of RNA degradation and therefore is used extensively in hybridizations involving this nucleic acid.

h. Reopen the bag and remove the probe solution as in step 2e.

h. The probe may be used for subsequent hybridizations if stored at −70°C.

i. Remove the nitrocellulose paper from the plastic bag and place it in a water-tight plastic container filled with $2\times$ SSPE containing 0.5% SDS (total volume 300 mL). Cover and incubate the filter at 70°C with gentle agitation for 15 minutes. Place a weight on the container to prevent it from floating. Change the wash solution and repeat the incubation and wash steps twice.

i. Because nitrocellulose tends to adhere to the plastic bag, open the bag by cutting three sides and peeling back the plastic. SDS blocks adsorption of single-stranded DNA and thereby facilitates removal of unbound probe.

j. Blot the nitrocellulose paper on filter paper and air dry. Place the nitrocellulose between two sheets of clear plastic.

k. Autoradiograph the nitrocellulose as outlined in Section XIII, p. 79. For ^{32}P-labeled probes use x-ray film, for ^{3}H-labels use fluorography.

k. Autoradiography with ^{32}P-labeled probes can be conducted within two to three hours (at room temperature) using an intensifier-lined exposure cassette. For ^{3}H-labeled probes 18−96 hours exposure time may be needed.

METHOD 2: COLONY OR PLAQUE HYBRIDIZATION

Colony hybridization was developed by Grunstein and Hogness (1975) as a modification of the Southern Blot procedure for use in screening of bacterial colonies for the carriage of a particular segment of DNA. Recently, modifications to the procedure have been developed to screen at high colony density (Hanahan and Meselson, 1983). A similar method was developed to screen for λ plaques containing cloned DNA (Benton and Davis, 1977). Plaque hybridization procedures have been of particular utility in screening GENE LIBRARIES. These procedures are widely used and may prove to be of particular utility in screening clinical isolates, such as enterotoxigenic *E. coli* (Moseley *et al.*, 1980, Georges *et al.*, 1983), for specific DNA sequences.

In the following, procedure A is a general method that combines aspects of both colony and plaque hybridizations. Procedure B is a recent modification that achieves increased sensitivity for the detection of single genes.

A. GENERAL HYBRIDIZATION METHOD

PROCEDURE

1. Carefully place a single thickness of circular nitrocellulose paper, 88 mm in diameter, over the discrete colonies or plaques appearing on the surface of a selective agar plate after overnight incubation at 37°C. For screening phage, plates are chilled at 4°C for 15 minutes to harden the top agar before placing the nitrocellulose paper on top.

2. Leave the filter paper on the agar surface for 1 to 20 minutes.

3. Remove the filter with sterile forceps and place it, colony side up, on several sheets of Whatman 3MM paper saturated with a solution of 0.5M sodium hydroxide and 1.5M sodium chloride (lysing/denaturing solution).

4. After 5 to 10 minutes, transfer the filter, colony side up, to dry 3MM paper to remove excess liquid.

5. Place the filter (colony side up) on 3MM paper saturated with neutralizing solution (1.5M NaCl, 0.5M Tris-HCl, pH 7.6) for 5 minutes.

6. Blot dry as indicated in step 4.

7. Air dry the filter (1 to 2 hours) and then bake *in vacuo* for 2 hours at 80°C.

8. The filters are ready for hybridization. The protocol for Southern Blots (Method 1 of this section, p. 60) should be followed with the following two modifications.

    a. Use 2 mL of prehybridization solution instead of 10 (step 2b).

    b. Use 2 mL of P^{32}-labeled probe instead of 5 mL (step 2f).

COMMENTS

1. Be sure to make an orientation mark on both filter and plates. This can be done with a marker pen on the plate and soft lead pencil on the filter paper. Use a glass hockey stick to ensure that no air bubbles are trapped between the agar surface and the filter paper. These interfere with transfer.

Whatman 541 high wet strength paper or nylon-based papers may be used instead of nitrocellulose for bacterial hybridizations (Gergen *et al.*, 1979).

2. Conditions of incubation may vary depending on the bacterial genus or phage. Longer periods are required for adsorbing virus to the filters.

3. If the cloning vector (plasmid) in the cell is amplifiable with chloramphenicol, put the filter paper, colony side up, onto an enriched agar plate containing 150 μg/mL chloramphenicol, and incubate overnight. Then proceed with the sodium hydroxide lysing step.

7. Nylon-based papers need not be baked *in vacuo*, while Whatman 541 paper requires no baking at all to fix the DNA.

PROCEDURE	COMMENTS
1. Carefully place an appropriately-sized piece of Whatman 541 paper over the bacterial colonies prepared as described in Step 1 of procedure A.	**1.** This procedure is modified from that of Maas (1983). Take care not to trap air between the filter and plate surface.
2. Leave the paper in place for 1–2 hours at room temperature.	
3. Peel off the 541 paper and place *colony side up* on a single sheet of Whatman 3MM paper that has been previously saturated with freshly prepared lysing/denaturing solution (step 3, procedure A) and placed in a heat-resistant dish.	**3.** Use sterile forceps to remove the paper.
4. Immediately after the 541 paper becomes thoroughly wet, place the dish over boiling water and steam for 3–5 minutes.	**4.** Pyrex ovenware is suitable for this. A glass beaker half filled with water heated over a Bunsen burner or hotplate serves as an adequate steamer. Steaming enhances the lysis of cells and the denaturation of their nucleic acid contents, thereby increasing the amount of DNA fixed to the filter paper.
5. Return the 541 paper to fresh lysing/denaturing solution at room temperature for 1 minute.	**5.** This permits cooling without renaturation.
6. Place the filter paper colony side up on 3MM paper to remove excess fluid.	**6.** This effects more efficient neutralization.
7. Place the filter (colony side up) on a piece of 3MM paper saturated with neutralizing solution (step 5, procedure A) for 5 minutes at room temperature.	
8. Blot the filter paper as in step 6 and leave to air dry at 37°C.	**8.** Unlike other hybridization papers, Whatman 541 does not require baking to fix DNA.
9. The filter is ready for hybridization as described in step 8 of procedure A.	**9.** Maas (1983) achieved good hybridizations when a ten-fold excess of probe DNA to fixed complementary DNA was used.

XII. Labeling DNA

DNA or RNA molecules displaying a high radioactivity per unit of mass (i.e., of high specific activity) are ideal probes for nucleic acid hybridizations. *In vivo* labeling of DNA by incubating whole cells in the presence of radioactive precursors is a rather inefficient way of producing such probes. However, a number of procedures, such as nick translation and end-labeling, are now available for producing highly labeled DNA molecules *in vitro*.

Cautionary Note: Handling of Radioactive Materials: Because the procedures that follow require the manipulation of radioactive materials, certain safety precautions are recommended: no mouth pipetting should be attempted, wear a lab coat and disposable plastic gloves, and conduct as many operations as possible over an area intended to contain spills (e.g., a leakproof tray lined with adsorbent paper). If the labeled material is in a volatile or powdered form, work in a fumehood. To minimize radiation exposure, all procedures involving high energy emitters (e.g., ^{32}P) should be carried out, if possible, behind a body shield constructed of 10 mm thick Plexiglas. In these instances spills can be detected by monitoring the working area with a Geiger counter.

Radioactive wastes should not be mixed with nonradioactive wastes but should be separated according to the isotope in use. Solid waste (gloves, tubes, paper, etc.) are collected in leakproof plastic bags (e.g., autoclave bags) and stored in an unbreakable container. Separate liquids into aqueous and compatable organic fluids. Water-based solutions, (e.g., column washings, Section XVII-N, p. 103; trichloro-acetic acid, Section XVII-O, p. 106), phenol, and ethanol, should be discarded into screw-capped, thick-walled glass containers; ether is discarded into a disposable beaker and allowed to evaporate in a fumehood. Tritium-labeled liquid should be mixed with an absorbent material such as kitty litter, then disposed of according to local regulations. In certain jurisdictions, low-level aqueous tritium solutions may be diluted with tap water, then washed down drains designated for this purpose. Because of its short half-life (14.3 days), ^{32}P-labeled material may be held in the liquid form for $3\frac{1}{2}-4$ months ($7\frac{1}{2}$ times half-life) before being disposed as nonradioactive material. Do not autoclave radioactive materials; biological contamination may be removed with appropriate dilutions of formalin. While radioactive wastes are being accumulated, they should be stored in a controlled access area away from other, general wastes.

A. NICK TRANSLATION

Nick translation refers to the limited digestion of purified DNA followed by the *in vitro* repair of the resultant lesions with a mixture of radioactive and cold deoxy-nucleoside triphosphates. This reaction is based on the activities of the enzymes DNase I and polymerase I. Pancreatic DNase I is used to introduce scissions or nicks into the DNA molecules. The exonuclease activity of polymerase I then acts at the sites of the nicks and removes nucleotides in the 5'→3' direction. At the same time, using the 3'-hydroxyl group of the terminal nucleotide of the nick as a primer, the polymerase activity of polymerase I will replace the pre-existing, unlabeled nucleo-tides with labeled nucleotides. As nucleotides are removed and added simultaneous-ly, the nick is moved or "translated" along the DNA molecule. Nick translation, which is used to prepare both isotope and biotin-labeled probes, results in uniform labeling throughout the DNA molecule. The specific radioactivity of the DNA is determined by the specific activity of the labeled nucleotides and by the extent of nucleotide replacement. The following procedure is a modification of the original method developed by Rigby *et al.* (1977).

PROCEDURE	COMMENTS
1. Dry (or concentrate) the labeling material.	**1.** ***CAUTION:*** *Because this procedure requires the manipulation of radioactive materials, the following precautions should be strictly observed: no mouth pipetting, wear disposable plastic gloves, conduct as many manipulations as possible over an area intended to contain spills. To minimize radiation exposure, all procedures involving radioactive materials should be carried out behind a body shield constructed of 10 mm thick Plexiglas. To detect spills, the working area should also be monitored with a Geiger counter.*

MATERIALS: Because DNA adheres to glass and some plastics, all procedures should be conducted using siliconized tubes and pipettes (see Section XVII-H, this chapter). To prolong their shelf life all reagents should be stored at −20°C. [α^{32}P] dNTP should be stored at −70°C. If nick translation kits are purchased commercially, the suppliers' instructions should be followed.

a. Place 5−50 μL of [α^{32}P] dCTP ($\simeq$ 400 Ci/mmol) or [methyl-1′, 2′−^{3}H] dTTP (100 Ci/mmol) in a 1.5 mL microfuge tube.

a. The labeling compound might be supplied in an ethanol solution. Because ethanol interferes with nick translation, it must be removed. Label may also be supplied in tricene buffer which does not interfere with the reaction. However, drying and resuspension of the labeled compound in a smaller volume (concentrated) is highly recommended, since one gains greater flexibility in the amounts of other reagents used, e.g., purified DNA. The amount of label used depends on the specific activity desired in the product, which in turn is determined by the half-life and the specific activity of the isotope.

Other labeled nucleotides may be added to increase specific activity.

b. Cover the tube with parafilm, and puncture the parafilm with a needle (5−6 holes).

c. Place the tube in a vacuum desiccator and dry at room temperature. Use immediately.

d. While the tracer is drying, bring the components of the reaction mixture to 0−4°C by placing in an ice bath.

2a. The nick translation reaction mixture is prepared on ice by adding the components listed below to the tube containing the labeled compound in the following order (final volume = 25 μL).

2a. Kits containing "ready-made" stock solutions are available commercially. If using pure enzymes from different suppliers, preliminary experiments may be necessary to establish optimal concentrations.

   i. 2.5 μL of 10× nick translation buffer, pH 7.5.

   i. **10× Nick Translation Buffer**

0.5 Tris-HCl (pH 7.5)
0.1M magnesium chloride
OPTIONAL (for additional stability): 500 μg/mL bovine serum albumin (BSA)
10 mM dithiothreitol (DTT)

   ii. 2.5 μL of each of the other three unlabeled deoxynucleoside triphosphates.

   ii. Dissolve each dNTP in 50 mM Tris-HCl, pH 7.5 to a final concentration of 0.3 mM.

   iii. 12 μL (0.5 μg) of the purified DNA to be labeled.

   iii. The final concentration of DNA in the reaction mixture should be approximately 20 μg/mL.

   iv. 2 units ($\approx$ 2 μL) of *E. coli* DNA polymerase I.

   iv. In 50 mM Tris-HCl, pH 7.5.

   v. 1 μL diluted DNase I solution (to start the reaction).

   v. ***DNase STOCK SOLUTION:*** 1 mg/mL DNase I in 50 μg/mL BSA plus (optional) 1 mM

DTT, 50% glycerol in 50 mM Tris-HCl, pH 7.5. Dilute 1:50,000 with cold buffer (minus glycerol) to generate probes less than 5 kb; 1:500,000 to generate duplex lengths of 20 kb. Stock solution should be aliquoted, then stored at $-20°C$. Use of a separate aliquot for each reaction will lessen the effect of repeated freezing and thawing on enzyme activity.

b. Cap the reaction tube and incubate at $12-14°C$ for 2−3 hours.

b. The time course of incorporation varies according to the activity of the individual components. Stop the reaction as it begins to plateau (this may require a trial run).

c. Stop the reaction by adding 25 µL 100 mM EDTA.

c. If labeled molecules are to be *immediately* separated from unincorporated precursor materials using a G-50 column, this step may be omitted.

3. Recover the labeled DNA by passing the reaction mixture through a Sephadex G-50 column (1 × 17 cm).

3. Prior to its use as a hybridization probe, nick translated DNA must be separated from unincorporated radioisotope; otherwise, troublesome levels of background radioactivity will occur. Particulars of the G-50 column are given in Section XVII-N, p. 103.

a. Elute the reaction mixture using TE buffer, pH 7.5. Collect two fractions of 1 mL volume, then 5−6 fractions of 0.5 mL volume. The labeled DNA is usually found in fractions 4 and 5.

a. Labeled DNA migrates approximately twice as rapidly as unincorporated triphosphates thereby giving good separation. With ^{32}P label, the progress of the DNA as it runs down the column can be monitored using a Geiger counter.

Instead of using a G-50 column, labeled DNA can be separated from unincorporated radioisotope by adding ammonium acetate (final concentration 0.3M) to the reaction mixture. Then the labeled DNA is precipitated with 2 volumes of cold ethanol. Usually, the recovery of labeled DNA is low. The addition of carrier DNA or RNA increases the efficiency of the precipitation of labeled DNA, thus the recovery.

b. Pool the peak fractions into a preweighed plastic vial. The weight of the vial contents is taken as an estimate of the volume of radioactive solution. Adjust the volume to 5 mL with TE buffer, pH 7.5.

b. If using ^{32}P-label, pass a Geiger counter along the collection tubes to determine peak fractions. If using ^{3}H-label, acid-fix 5 µL aliquots of each fraction, on the filter and detect the peak by liquid scintillation counting (see Section XVII-O, p. 106).

^{32}P-labeled probes should have at least 10^6-10^7 counts per minute per mL.

2. Use as probe for hybridization as outlined in Section XI, p. 59.

B. END-LABELING

Hybridization probes prepared by nick translation are uniformly labeled along the length of the DNA. However, other techniques, such as nucleotide-sequencing using Maxam and Gilbert's method (1980; Section XVI, p. 94), or restriction endonuclease mapping by partial digestion (Section IX, p. 35), require the DNA molecule to be labeled at its termini. Depending on the procedure used, linear DNA fragments can be radioactively labeled at either their 3' or 5' ends. Unlike nick translation, end labeling requires the linearization of circular DNAs, thus the restriction endonuclease susceptibility of the DNA molecules must be known.

METHOD 1: 3'-END-LABELING WITH *E. COLI* DNA POLYMERASE I KLENOW FRAGMENT OR T4 DNA POLYMERASE (FILLING-IN OR REPLACEMENT SYNTHESIS REACTIONS)

The 3'-end can be labeled by using the Klenow fragment of *E. coli* polymerase I (Drouin, 1980), T4 DNA polymerase (O'Farrell *et al.*, 1980; Challberg and Englund, 1980) or terminal deoxynucleotidyl transferase (Roychaudhury and Kossel, 1971). The Klenow fragment is a 76,000 dalton polypeptide that is released from DNA polymerase I after treatment with subtilisin (Jacobsen *et al.*, 1974). This novel enzyme possesses the 5'→3' polymerase and 3'→5' exonuclease activities, but not the 5'→3' exonuclease activity of DNA polymerase I. Depending on reaction conditions, both the Klenow fragment and T4 polymerase are able to mediate 3'-end-labeling in two ways: the filling-in reaction and the replacement synthesis reaction. Both reactions require the presence of a 5'-overhang in the substrate DNA. These are generated in a number of ways: 1) by digestion with restriction endonucleases generating 5'-tails; 2) by the addition of linker DNA with recognition sites for restriction endonucleases generating 5'-overhangs; 3) by Klenow fragment or T4 polymerase (3'→5' exonuclease activity) digestion of DNA preparations having either blunt ends or 3'-overhangs. Since T4 DNA polymerase has the higher exonuclease activity, it is the enzyme of choice for this procedure (Maniatis *et al.*, 1982).

Filling-in Reaction: All four deoxyribonucleoside triphosphates, one of which is labeled, are included in the reaction mixture. The 5'→3' polymerase activity of the DNA polymerase simply extends the recessed 3'-strand using the protruding 5'-overhang as a template. Filling-in synthesis results in DNA fragments with blunt ends as indicated below.

$$
\begin{array}{lll}
 & [\alpha^{32}\text{P}]\text{dC*TP} & \\
\text{3'-G-A-A-T-T-G-A-T-C- 5'} & \xrightarrow{} & \text{3'-G-A-A-T-T-G-A-T-G- 5'} \\
\text{5'-C-T-T-A-A} \qquad\quad \text{3'} & \text{dATP} & \text{5'-C-T-T-A-A-C*-T-A-G- 3'} \\
 & \text{dGTP} & \\
 & \text{dTTP} & \\
 & +\ \text{Klenow fragment} &
\end{array}
$$

Replacement Synthesis Reaction: In the replacement reaction, only one labeled deoxyribonucleoside triphosphate is included in the reaction mixture. The three other dNTPs are excluded. Thus, only one nucleotide per end is labeled in the reaction. For example (Figure 1.12a and b), when $[\alpha^{32}\text{P}]$ dTTP is used to label DNA molecules with either 5'-protruding or blunt ends, the 3'-terminal dTMP is removed through the 3'→5' exonuclease activity of the DNA polymerase. However, before additional nucleotides are removed, the 5'→3' polymerase activity of the enzyme predominates and the terminal dTMP is replaced by transfer from $[\alpha^{32}\text{P}]$ dTTP.

If, on the other hand $[\alpha^{32}\text{P}]$ dCTP was used the sole nucleoside triphosphate instead of $[\alpha^{32}\text{P}]$ dTTP, the polymerase I enzyme would remove three nucleosides from the fragment illustrated in Figure 1.12a and four nucleosides from the fragment illustrated in Figure 1.12b before replacing the dCMP by transfer from $[\alpha^{32}\text{P}]$ dCTP. Thus, in these examples the labeled strands are shortened by 2 and 3 nucleosides, respectively. This is important to remember when sequencing procedures are carried out.

a. MOLECULES WITH 5′-PROTRUDING ENDS

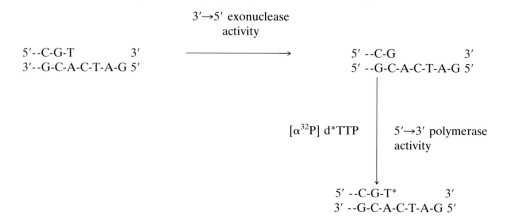

b. MOLECULES WITH BLUNT ENDS

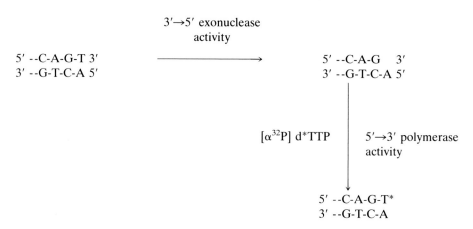

FIGURE 1.12 (*a and b*)   3′-End-labeling replacement synthesis reaction.

PROCEDURE	COMMENTS

1. Filling-in Reaction

Place the following reagents in a 1.5 mL microfuge tube:

a.   0.05 mCi [α^{32}P] dTTP (dry).

a.   If the [α^{32}P] dNTP is in a solution containing ethanol, it is necessary to lyophilize it before adding other ingredients (as in the nick translation reaction).

b.   8 µL of DNA (final concentration 30−50 µg/mL) solution (fragment in H$_2$O).

b.   3′-end labeling incorporates radioactive isotopes only at the ends of the DNA fragments. In order to achieve high specific activity, a high concentration of DNA fragment is required, so as to provide a large number of ends for labeling. Less DNA is needed to achieve this with short DNA fragments than for long fragments.

c.   5 µL reaction buffer (10× nick translation buffer).

c.   Refer to part A, p. 69, this section, for buffer recipe.

d. 2 μL (2 units) Klenow fragment of DNA polymerase I.

d. When labeling DNA fragments with blunt or protruding 3'-ends, the reaction mixture should be pre-incubated at 15°C for 30 minutes. This allows the DNA polymerases to perform their 3'→5' exonuclease digestion before being overwhelmed by their polymerase activity.

e. 5 μL of unlabeled dNTP mixture (solution containing 0.1 mM each of the 3 unlabeled deoxyribonucleoside triphosphates).

e. ***REPLACEMENT REACTION:*** The conditions for replacement reaction are the same as those for the filling-in reaction except that the mixture of non-radioactive dNTP is not added.

2. Mix and incubate at 15°C for 1−2 hours.

3. Stop the reaction with 5 μL of 0.1M EDTA.

3. This step may be omitted, as subsequent treatment with phenol will stop the reaction.

If concentrated DNA is not required, the reaction mixture can be loaded directly onto a Sephadex G-50 column to separate unincorporated [α^{32}P] dNTP (see Section XVII-N, p. 103).

4. Add 100 μL of 0.3M sodium acetate to the solution, and mix.

5. Add an equal volume of saturated phenol, mix by inversion and centrifuge to separate the phases.

5. Phenol is used to extract the DNA polymerase I (Klenow fragment.) Sometimes it is hard to remove the aqueous phase especially when the volume is small. In these cases, bring the volume up to 500 μL with 0.3M sodium acetate before phenol extraction.

6. Remove the upper aqueous layer to a new 1.5 mL microfuge tube and add 500 μL ether. Mix and centrifuge. Remove the upper organic layer containing any residual phenol. Repeat this step once more.

6. Ether is used to extract any residual phenol present in the aqueous solution.

7. Add two volumes of ice-cold ethanol and mix gently. Chill −70°C for 15 minutes or −20°C for 2 hours.

8. Pellet the DNA in a microfuge for 5−15 minutes.

9. Remove the supernatant and dry the pellet, either under vacuum or by inverting the tube on a rack (air dry).

10. Resuspend the labeled DNA in an appropriate buffer.

10. The labeled DNA can be further purified by gel electrophoresis and subsequent extraction from the gel (Sections V, p. 18, and VIII, p. 30).

METHOD 2: 3'-END-LABELING WITH TERMINAL DEOXYNUCLEOTIDYLTRANSFERASE

Terminal deoxynucleotidyltransferase (terminal transferase) is a small basic protein of molecular weight 33,000, which behaves like a DNA polymerase in synthesizing a DNA chain by 5'→3' polymerization of 5'-deoxyribonucleoside triphosphates. However, unlike DNA polymerase, it does not require a template.

Roychaudhury and Kossel (1971) first described a procedure that used terminal deoxynucleotidyltransferase to label the 3'-end of DNA with radioactive ribonucleoside triphosphates. This procedure was subsequently improved by Roychaudhury and Wu (1980). They observed that by replacing Mg++ with Co++, terminal deoxynucleotidyltransferase can accept double-stranded DNA as primer and label the 3'-ends of the DNA molecule whether it has blunt ends or protruding 3'-ends. When the reaction is extensive, chain elongation occurs. Tu and Cohen (1980) observed that terminal deoxynucleotidyltransferase can catalyse the addition of the nucleotide analog, cordycepin 5'-triphosphate (3'-deoxyadenosine 5'-triphosphate), to the 3'-ends of a DNA molecule. Since cordycepin 5'-triphosphate does not have a 3'-hydroxyl group, further chain elongation is thus prevented by the addition of a single cordycepin monophosphate.

PROCEDURE

1. Place 0.1 mCi [α^{32}P] cordycepin 5'-triphoshate in a 1.5 mL microfuge tube. Lyophilize if necessary.

2. Add the following to the microfuge tube:

a. 25 μL of DNA ($\approx$1 μg) in H_2O.

b. 3 μL of 10$\times$ tailing buffer.

c. 2 μL (10$-$15 units) of terminal transferase.

3. Incubate for 30$-$60 minutes at 37°C.

4. Add 400 μL of 0.3M sodium acetate. Add an equal volume of saturated phenol, mix by inversion and centrifuge to separate the phases. Remove the upper aqueous layer to a new 1.5 mL microfuge tube and add 500 μL ether. Mix and centrifuge. Remove the upper organic layer containing any residual phenol. Add two volumes of ice-cold ethanol and mix gently. Chill at -70°C for 15 minutes or -20°C for 2 hours. Pellet the DNA by centrifugation at full speed in a microfuge for 5$-$15 minutes. Remove the supernatant and dry the pellet either under vacuum or by inverting the tube on a rack (air dry).

5. The labeled DNA can be separated from unincorporated labels by gel electrophoresis or Sephadex G-50 column (Section XVII-N, p. 103).

COMMENTS

1. Lyophilize the radioisotope if it is suspended in ethanol or if the volume required is too large (>25 μL).

a. The amount of DNA used depends on its size. The larger the DNA fragment, the greater the amount required to give the same number of termini as smaller DNA fragments.

b. **10$\times$ Tailing Buffer**

   1 M cacodylic acid, sodium salt
250 mM Tris-base (adjust to pH 7.6 with KOH)
   2 mM dithiothreitol
  10 mM cobalt chloride

(To avoid precipitation, 0.1M cobalt chloride solution is added drop by drop to the buffer with constant stirring).

METHOD 3: 5'-END-LABELING

Two types of reactions, both involving polynucleotide kinase and $[\gamma^{32}P]$ ATP, are used to label the 5'-ends of DNA molecules (Chaconas and Van de Sande, 1980). In the first reaction, the forward reaction, the terminal 5'-phosphate of the DNA to be labeled is dephosphorylated (by alkaline phosphatase); then, the labeled gamma phosphate of $[\gamma^{32}P]$ is transferred to the resultant 5'-hydroxyl ends with polynucleotide kinase. In the second reaction, the reverse or exchange reaction, polynucleotide kinase catalyses the direct exchange of the gamma phosphate of $[\gamma^{32}P]$ATP with the terminal 5'-phosphate in the presence of excess $[\gamma^{32}P]$ATP and ADP. Generally, the exchange reaction is not as efficient as the forward.

a. FORWARD REACTION

PROCEDURE	COMMENTS
1. To dephosphorylate the DNA, add the following into a 1.5 mL microfuge tube and mix.	**1.** Either bacterial alkaline phosphatase or calf intestinal alkaline phosphatase can be used. The latter has about 10- to 20-fold higher activity per mg of protein than the bacterial enzyme (Chaconas and Van de Sande, 1980).
a. 5 μL (approximately 1 μg) of DNA fragment.	
b. 25 μL of 50 mM Tris-HCl (pH 8.0) buffer.	
c. 20 μL (2−5 unit/mL) alkaline phosphatase.	
d. Incubate at 37°C for 30−60 minutes.	
e. Denature and remove alkaline phosphatase by phenol extraction followed by ether extraction. Add 3.0M sodium acetate to a final concentration of 0.3M and precipitate with ethanol.	*e.* Alkaline phosphatase may also be inactivated by treatment with NaOH or HCl.
2. Lyophilize 0.1 mCi $[\gamma^{32}P]$ ATP in a 1.5 mL microfuge tube.	**2.** Lyophilization is necessary, as $[\gamma^{32}P]$ ATP is usually suspended in ethanol which may interfere with subsequent reactions.
3. Add the following components to the same tube:	**3.** Polynucleotide kinase catalyzes the transfer of the γ-phosphate of ATP to the 5'-OH terminus of the DNA.
a. 38 μL (≈ 1 μg) of dephosphorylated DNA fragment.	
b. 10 μL of 5× forward reaction buffer.	*b.* **5× Forward Reaction Buffer** 500 mM Tris-HCl, pH 7.6 100 mM magnesium chloride 50 mM dithiothreitol 1 mM spermidine
c. 2 μL (2 units) of polynucelotide kinase.	
d. Incubate at 37°C for 30−40 minutes.	
e. Add an equal volume of saturated phenol, mix by inversion and centrifuge to separate the phases. Remove the upper aqueous layer to a new 1.5 mL microfuge tube, and add 500 μL ether. Mix and centrifuge. Remove the upper organic layer containing any residual phenol. Add 3.0M sodium acetate to a final concentration of 0.3M.	

Add two volumes of ice-cold ethanol and mix gently. Chill at −70°C for 15 minutes or −20°C for 2 hours. Pellet the DNA by centrifugation at full speed in a microfuge for 5−15 minutes. Remove the supernatant and dry the pellet either under vacuum or by inverting the tube on a rack (air dry).

b. REVERSE (EXCHANGE) REACTION

1. Lyophilize 0.1 mCi of $[\gamma^{32}P]$ ATP as outlined in the forward reaction.

2. To the above add the following:

a. 38 μL (approximately 1μg) of DNA in H_2O.

b. 10 μL of 5× reverse reaction buffer.

c. 2 units of polynucleotide kinase.

3. Incubate at 37°C for 30−60 minutes.

4. Phenol and ether extract, add sodium acetate to a final concentration of 0.3M and precipitate the labeled DNA as outlined in step 2e of the 5′-end-labeling forward reaction.

1. Labeling 5′-ends using the exchange reaction does not require a dephosphorylated DNA substrate. The procedure is very similar to that of the forward reaction.

b. **5× Reverse (Exchange) Reaction Buffer**

500 mM imidazole-HCl, pH 6.6
100 mM magnesium chloride
 50 mM dithiothreitol
  1 mM spermidine
  3 mM adenosine diphosphate (ADP)

C. NON-RADIOACTIVE PROBES (BIOTIN-AVIDIN)

Langer-Safer *et al.* (1982) have described a hybridization procedure that employs an immunological rather than a radioisotopically-labeled probe. DNA containing biotinylated deoxyuridine triphosphate (UTP), incorporated by nick translation, is hybridized with the test nucleic acid (either *in situ* or immobilized on nitrocellulose paper). The biotin molecules in the probe serve as antigens which bind rabbit antibiotin antibodies. The hybridization sites are subsequently detected either fluorimetrically with fluorescein-labeled anti-rabbit IgG, or colorimetrically with anti-rabbit IgG antibody conjugated to horseradish peroxidase. The latter assay is most useful for monitoring membrane-based hybridizations. The sensitivity of this procedure has been improved by the development of a colorimetric reagent based on complexes of the glycoprotein, avidin (or streptavidin), and biotinylated polymers of alkaline phosphatase (Leary *et al.*, 1983). With this reagent, such enzyme-amplified immunoassays can be used to visualize specific DNA sequences in the 1−10 picogram (pg) range in less than one hour.

Recently, a method for directly labeling DNA with colorimetrically detectable enzymes has been developed (Renz and Kurz, 1984). The procedure, which may replace biotin-avidin systems, detects sequences a few kilobases long. These approaches offer greater probe shelf-life, less background, and equal or better resolution power than the isotopic methods. They will be of special significance in laboratories not equipped to handle radioisotopes.

The following section outlines a labeling and hybridization procedure adapted for use with such biotin-avidin probes. It is based on the instructions provided with the Enzo Bio-Probe™ System produced by Enzo Biochem Inc., New York. It should be noted that other companies (Bethesda Research Laboratories) also produce reagents for detecting biotinylated DNA.

PROCEDURE

1. The DNA is labeled by nick translation (see part A, p. 68, of this section for an introduction to the principles of this reaction).

a. Remove the kit from storage at −20°C and warm to 0°C.

b. Add each of the following components in the order and amounts indicated to a fresh 1.5 mL microfuge tube, standing on an ice bath.

    i. 5 μL of 10× biotin-Nick translation buffer

    ii. 5 μL dNTP solution

    iii. 5 μL biotinylated dUTP

    iv. 1−5 μg of DNA to be labeled

    v. Sufficient distilled H_2O to give a final volume of 50 μL

    vi. 4 μL DNA polymerase I (3 units/μL)

COMMENTS

a. Place each component in an ice bath to "warm" it.

    i. **10× Biotin-Nick Translation Buffer**

0.5M Tris-HCl, pH 7.5
50 mM $MgCl_2$

    ii. **dNTP Solution**

0.30 mM dATP
0.30 mM dGTP
0.30 mM dCTP
50 mM Tris-HCl, pH 7.5

    iii. **Biotinylated dUTP**

0.30 mM biotinylated dUTP
50 mM Tris-HCl, pH 7.5

Biotin-dUTP serves as an dTTP analog.

    iv. Purified as outlined in Sections I and II of this chapter. Lyophilization may be used to concentrate the DNA.

    v. Amount used will be determined by the DNA concentration.

    vi. DNA polymerase I is suspended in phosphate buffer (pH 7.2).

Phosphate Buffer

0.1M sodium phosphate
50% glycerol
1.0 mM dithiothreitol

PROCEDURE	COMMENTS

vii. 4 μL DNase I (0.1 μg/μL)

vii. Dilute 0.5 mg/mL stock of DNase I solution 5000-fold by: 1) diluting 1:5 with DNase buffer; 2) diluting an appropriate amount of the resultant solution 1:1000 with distilled H_2O.

This concentration (400 pg per reaction mixture) yields λ DNA duplexes 0.8–1.0 kb in length.

DNase Buffer

10 mM Tris-HCl, pH 7.5
1 mg/mL bovine serum albumin (BSA)

c. Incubate the reaction mixture at 14°C for 2 hours.

c. The inclusion of ^{3}H-dATP (10–12 μL of 0.25 microCi/μL stock; lyophilize to remove ethanol) in the reaction mixture enables the kinetics of incorporation to be monitored. Substitution of 20-40% of the unlabeled dNTP by labeled nucleotides is obtained on average.

d. Stop the reaction by adding 5 μL of 0.2M EDTA. Heating at 65°C for 10 minutes.

d. Alternatively, labeled DNA may be separated from unincorporated dNTP's on a Sephadex G-50 column. See Section XVII-N, p. 103, for Sephadex G-50 chromatography.

2. The probe is hybridized to immobilized DNA.

a. Denature and transfer the DNA to nitrocellulose paper as outlined in Method 1, Section XI, p. 60.

a. Blotting papers and transfer methods may be varied according to preference and requirements. Avidin-biotin probes are also useful in dot-blots.

b. Hybridize with the avidin-biotin labeled probes as outlined in Method 1, Section XI, p. 60. In this instance use 2× SSC plus 0.1% SDS *in place of* 2× SSPE + 0.5% SDS for the final wash. For the last wash of the series, eliminate the SDS and carry out at room temperature.

b. The manufacturer recommends the use of 400–500 ng/mL of biotinylated DNA per hybridization. Unlike radioisotopically-labeled probes, a high concentration of biotin probe can be used to achieve more efficient detection in a shorter time without producing high backgrounds.

3. The hybridized biotinylated DNA is detected.

a. **10× Phosphate Buffered Saline (PBS Stock)**

1.3M sodium chloride
0.7M sodium phosphate (dibasic: Na_2HPO_4)
0.03M sodium phosphate (monobasic: NaH_2PO_4)

a. Immerse the still-wet filter paper from the final hybridization wash in 1× phosphate buffered saline containing 2% BSA and 0.1% Triton X-100.

Triton X is a nonionic surfactant. BSA, or any other inert protein, inactivates or "blocks" unused protein-binding sites.

b. Soak the filter for 30 minutes at 37°C, then discard the buffer.

b. This step increases the efficiency of subsequent detection steps by minimizing non-specific interactions.

c. Immerse the filter paper in horseradish peroxidase (hrp) detection complex (use 0.02 mL/cm^2) for 30 minutes at 37°C.

c. HORSERADISH PEROXIDASE DETECTION COMPLEX: dilute streptoavidin-biotinylated horseradish peroxidase complex 1:250 with 1× PBS containing 1% BSA. This reagent acts as antibody to antigenic biotinylated probe hybridized to test DNA. Minimize background by conducting the detection steps in the same container as used for the blocking.

*The alkalaline phosphatase detection system is about 10 times more sensitive.

d. Discard the hrp complex and wash the paper three times at 5 minute intervals with high-salt buffer.

d. High Salt Buffer

  0.5M sodium chloride
10 mM phosphate buffer, pH 6.5
  0.1% BSA
Tween 20 (0.5 mL/L)

e. Wash twice with 2× SSC (see Method 1, Section XI, p. 60, for SSC formula) containing 0.1% BSA and Tween 20 (0.5 mL/L).

e. The washings remove nonspecifically-bound hrp complex (no antigen/antibody interaction).

f. Assay for peroxidase activity by applying visualization solution (0.02 mL/cm^2) to the hrp-treated paper.

If present, a purple-blue color will appear in 2–10 minutes. If colony hybridization procedures were used, the colony will not be uniformly stained; rather, a purple-brown ring will form at the perimeter of the colony.

f. Visualization Solution

0.5 ng/mL diaminobenzidine tetrahydrochloride (DAB)
10 mM Tris-HCl, pH 7.5
0.02% cobalt chloride
0.3% hydrogen peroxide

Prepare the DAB portion fresh daily. Use 1 and 30% stock solutions of cobalt chloride and hydrogen peroxide, respectively. Add cobalt chloride to DAB solution, mix well and place on ice in the dark for 10 minutes. Lastly, add hydrogen peroxide. The entire mixture is photosensitive.

The nitro groups present on the nitrocellulose membrane give it a strong oxidizing potential. It is therefore possible to get false positives during adsorption of the visualization solution. Appropriate controls should be performed. Alternatively, the DAB may be replaced by a substrate stain such as 4-chloro-1-naphthol which is resistant to such enzyme-independent oxidation.

XIII. Autoradiography and X-Ray Film Development

Autoradiography, a technique based on film imaging methods, is used to detect radioisotopically-labeled materials. An x-ray film is directly exposed to radioactive compounds present in electrophoresis gels, or other materials. Radioactive decay products strike the film interacting with the silver halide contained in the emulsion, thereby producing an (latent) image. This image can then be visualized using standard photographic development procedures. High energy gamma-emitters, such as ^{32}P or ^{125}I, can be detected directly by this technique. Visualization of low energy beta-emitters such as ^{3}H, ^{14}C or ^{35}S requires an intermediate step whereby emitted energy is converted to more readily detected light energy (fluorographic enhancement). The organic scintillator, PPO (2,5-diphenyloxazole), may be used for this purpose (Bonner and Laskey, 1974).

METHOD 1A: AUTORADIOGRAPHY

NOTE: *All steps should be carried out in a photographic darkroom.*

PROCEDURE	**COMMENTS**

<table>
<tr><td></td><td>Store unopened boxes of film at 4°C, and open after bringing to room temperature. Wrap opened boxes in a light-impervious material, such as black plastic garbage bags.</td></tr>
<tr><td>**1.** With the safety light (red filter) on, place one sheet of 20.3 × 25.4 cm Kodak X-Omat RP (code XRP-1) film next to the hybridization paper or gel which has been enclosed in plastic as described in Section XI, Method 1, p. 60.</td><td>**1.** A standard frosted 15–20W bulb masked by a Kodak Type GBX-2 (red) filter will serve as a safety lamp. X-ray film can be hypersensitized by exposure to a single instantaneous flash of light, immediately before use. However, the length of exposure must be very short (in the order of 1 mSec.). Without proper equipment, it is NOT advisable to perform such a procedure, as the longer exposure will increase the "fog" level instead of hypersensitizing the film. Other film types which will give good results include: Kodak X-Omat AR film or Kodak SO-445 film. Both are of higher speed than the XRP-1. The faster films are appropriate for fluorography.</td></tr>
<tr><td></td><td>Increased resolution is achieved if the gel is dried prior to exposure. Slab gel driers are available commercially.</td></tr>
<tr><td>**2.** Use scissors to cut an orientation mark in the two sheets.</td><td>**2.** Alternatively, a small amount of radioactive label may be spotted on one corner of the paper.</td></tr>
<tr><td>**3.** Place the film and filter paper back-to-back in an X-ray exposure cassette that is lined with image-intensifying screens.</td><td>**3.** Not only does the x-ray cassette protect both film and worker, but it also assures that the entire surface of the filter paper is pressed tightly against the film, thereby preventing loss of resolution. **Image intensifying screens**, containing gamma-activated inorganic phosphorus, reduce exposure times by bouncing detectable light back to the film.</td></tr>
<tr><td>**4.** Expose the film for 2–3 hours at room temperature.</td><td>**4.** Exposure times are determined by the type and amount of radioisotope used for labeling. For particularly hot samples adequate exposure may be achieved within one-half hour. If sufficient space is available, film sensitivity and image formation may be maximized by placing the loaded cassette at −70°C for the exposure period. If these conditions are used, the cassette must equilibrate to room temperature before the film is retrieved and processed, or some reduction in image quality may occur.</td></tr>
</table>

METHOD 1B: FLUOROGRAPHY

PROCEDURE	**COMMENTS**
1. Immerse nitrocellulose hybridization membranes in a 20% (w/v) solution of 2,5-diphenyloxazole (PPO) in toluene for 5–10 minutes.	**1.** PPO is an organic scintillator that converts β-energy to light energy.
	The membrane must be thoroughly dried, otherwise the PPO will not be efficiently adsorbed (Laskey, 1980).

PROCEDURE	COMMENTS

	Alternatively, the filter may be coated with a commercial fluorographic preparation (autoradiography "enhancer") available as an aerosol spray.
2. Air dry, then process as described in the previous method.	**2.** Expose the film for 18–24 hours.
	If the separation gel itself is to be exposed to X-ray film, it must first be impregnated with PPO dissolved in dimethylsulphoxide (DMSO), or sodium salicylate in water, then extensively washed and dried prior to being processed as in Method 1A, p. 79. Both polyacrylamide and agarose gels may be treated with salicylate, whereas the DMSO must only be used with polyacrylamide (see Grierson, 1982, for details).

METHOD 2: DEVELOPING X-RAY FILM

PROCEDURE	COMMENTS
	CAUTION: Turn on the safety light to remove the exposed X-ray film from its cassette.
1. Wash the film for 3 minutes in Kodak liquid X-ray developer.	**1.** **Plastic gloves should be worn.** Use photographic tongs to immerse and transfer the film. The times indicated are for freshly prepared solutions. Replace developer when it turns brown.
2. Immerse in stop solution (3–5% acetic acid) for 30 seconds.	
3. Wash for 3 minutes in Kodak fixer.	**3.** After a few seconds in the fixer, normal lighting may be used.
4. Rinse the film under running water for 10–30 minutes.	**4.** Do not use warm water, as the image on the film may peel off.
5. Hang the film to dry.	

XIV. Transformation of *Enterobacteriaceae*

Transformation describes the processes involved in the uptake, integration and expression of naked DNA from the surrounding medium by competent cells. The uptake of naked viral or plasmid DNA by cells has been termed transfection (for a review see Smith and Danner, 1981), although transformation is the more colloquial usage in the case of plasmid DNA. Many genera of bacteria are naturally competent; these include *Bacillus, Streptococcus, Haemophilus, Moraxella, Acinetobacter* and *Neisseria* species. Other species, such as *E. coli*, are not naturally competent and until recently, could not be transformed. Similarly, naturally competent bacteria are either inefficiently or not transfected by phage and plasmid DNAs (Smith and Danner, 1981). However, techniques have been developed which enable some noncompetent bacteria to be transformed or transfected. These technological advances have enabled genetic exchange to take place between completely unrelated organisms.

Mandel and Higa (1970) found that *E. coli* cells could be transfected with phages P2 and λ after treatment with calcium ions. The method was modified by Cohen *et al.* (1972) for the uptake of plasmid DNA by CaCl$_2$-treated *E. coli*. Subsequently, modified procedures for creating competent bacteria by "artificial" means have been developed for such genera as *Salmonella* (Lederberg and Cohen, 1974) and *Pseudomonas* (Chakrabarty *et al.*, 1975; Mercer and Loutit, 1979).

Efficiencies of transformation or transfection are influenced by a variety of conditions including the kind and concentration of divalent cation used; the concentrations of host cells and DNA; the genetic state of the recipient cell—certain mutants that have altered cell envelopes or that do not possess restriction-modification systems give higher efficiencies of transformation; and other factors such as pH, length of incubation of DNA/cell mixtures, length of heat shock, and plating conditions.

The following is a general procedure which gives good transformation efficiencies (10^5 to 10^6 transformants per μg of DNA) with a variety of small (4.2 to 16 kb) plasmids from several species. It is derived from methods published by Cohen *et al.* (1972) and Lederberg and Cohen (1974).

PROCEDURE

1. Dilute an O/N L broth culture of the recipient bacteria 1:100 in 50 mL of L broth and grow to mid-log phase (approximately 5×10^8 cells) in a 37°C shaking water bath (approximately 90 to 120 minutes).

2. Chill the cells in an ice bath for 15 minutes. Pellet by centrifugation at $7,000 \times g$ for 10 minutes at 4°C. Decant the supernatant.

3. Resuspend the pellet in 50 mL of cold 0.1M MgCl$_2$. Centrifuge and discard the supernatant as described in step 2.

4. Resuspend the pellet in 25 mL of cold 0.1M CaCl$_2$. Chill for 20 minutes in an ice bath.

COMMENTS

1. This transformation procedure achieves good results with *E. coli* strains HB101, C600 or 490; however, it must be remembered that different strains have different transformation efficiencies. Modified procedures are recommended for transforming *E. coli* χ1776 (Norgard *et al.*, 1978; Bolivar and Backman, 1979).

L (Luria) Broth:

    10 g Bacto tryptone
     5 g Bacto yeast extract
    0.5 g NaCl
     1 L dH$_2$O

Adjust to pH 7.0 with 1M NaOH. Add 10 mL of 20% glucose after autoclaving.

2. It is important that cells be kept cold. To maintain temperature, precool pipettes and tubes to 4°C.

3. Competent cells are fragile. Do not vortex to resuspend. MgCl$_2$ may be replaced with 10 mM NaCl. Remember to sterilize both the MgCl$_2$ and CaCl$_2$ solutions.

Several workers have obtained increased efficiencies of transformation by treating cells with a combination of calcium and rubidium chloride (see Kushner, 1978 and Norgard *et al.*, 1978).

PROCEDURE	COMMENTS

5. Pellet cells as before. Resuspend the pellet in 2.5 mL cold 0.1M CaCl₂.

5. At this step, several options are available (note that some of the options may reduce the frequency of transformation).

    a. Longer incubation in calcium chloride improves the competence of *E. coli* cells (Dagert and Ehrlich, 1979). Competent cells may be left in 0.1M CaCl₂ overnight at 0°C. They can then be spun down and resuspended in 2.5 mL of fresh, cold 0.1M CaCl₂ as described in step 5.

    b. Competent cells may be stored at −70°C or lower, by adding 15% glycerol to the CaCl₂ solution (Morrison, 1977). When required, cells are pelleted and resuspended in 2.5 mL cold 0.1M CaCl₂, as in step 5.

6. Add 0.1 mL (1−5 μg/mL) of chilled DNA solution to 0.2 mL of competent cells. Keep the cells on ice for 30 minutes.

6. DNA is prepared as described in Sections I and II. Incubation of cells at 0°C increases the frequency of transformation (Cohen *et al.*, 1972). It should be remembered that the efficiency of transformation decreases as the mass of the DNA increases.

7. Pulse-heat the transformation mixture by placing it at 42°C for 2 minutes.

7. Hot water from the tap can be used for this step. The heat pulse facilitates uptake of DNA by competent cells.

8. Chill the cells for 10 minutes in an ice bath.

9. Dilute the sample 1:10 into prewarmed Luria broth and incubate with moderate shaking for 90−120 minutes at 37°C.

9. This step ensures full expression of transforming DNA.

10. Plate appropriate dilutions of the cells on selective media and incubate overnight at 37°C.

10. Incorporate both positive (transformed with known entity) and negative (untransformed) controls. These controls test for competence and the frequency of spontaneous mutation, respectively.

XV. RNA Techniques

A. EXTRACTION AND ISOLATION OF EUKARYOTIC MESSENGER RNA (mRNA)

Many different mRNA's are present in various tissues, and have similar chemical and physical properties. The isolation of a specific mRNA can be achieved only by extraction from cells or tissues in which it shows an unusually high concentration, either naturally or after induction (for review, see Taylor, 1979). Total cellular RNA is usually extracted from tissue homogenates (Swan *et al.*, 1972) or cell lysates (Lee *et al.*, 1971a) with a detergent such as sodium dodecyl sulfate (SDS), and an organic solvent such as phenol (Lee *et al.*, 1971b; Kirby, 1968). Unlike prokaryotic mRNA, eukaryotic mRNA contains a covalently attached 3′-terminal polyadenosine [poly(A)] segment (tail) ranging in size from 20–250 bases (Molloy and Puckett, 1976), and thus can easily be separated from other RNA by affinity chromatography on oligodeoxythymidine (oligo(dT))-cellulose (Aviv and Leder, 1972), or polyuridine [poly(U)] agarose (Lindberg and Persson, 1972) columns. Specific mRNA can then be enriched and isolated by linear sucrose gradients, or electrophoresis on denaturing gels.

Successful RNA isolations are dependent on the inhibition or inactivation of ribonucleases present in the cell extract. For this purpose, inhibitors such as heparin (Zöllner and Felling, 1953), bentonite (Singer *et al.*, 1961), polyvinyl sulfate (Daigneault *et al.*, 1971), diethyl pyrocarbonate (Rosén and Fedorscak, 1966), or RNasin (Gagnon and de Lamirande, 1973) are often included in preparing tissue homogenates or cell lysates. Because they are strong denaturants of proteins, detergents, especially SDS, may be used as well. Since the type of ribonuclease present varies from tissue to tissue, combinations of inhibitors are often used to ensure complete inactivation.

The following is a general procedure modified from Lee *et al.* (1971a and b) and Aviv and Leder (1972) for the preparation of eukaryotic mRNA.

PROCEDURE

1. Suspend 50 g of sliced tissue in 125 mL (2.5× volume of tissue) of homogenizing buffer containing 500 μg/mL of polyvinyl sulfate and 5 mg/mL of bentonite. Homogenize the tissue in a Dounce homogenizer or a Waring Blender for 30–60 seconds.

2. Centrifuge the homogenate at 12,000 × *g* for 15–20 min at 4°C. Transfer the supernatant to a fresh centrifuge tube.

3. Centrifuge at 100,000 × *g* for 3 hours at 4°C in order to pellet the polysomes. Decant the supernatant.

4. Resuspend the pelleted polysomes in 2 mL of extraction buffer.

5. Add 0.2 mL of 10% SDS (in extraction buffer) to the polysome suspension (final concentration is 1%).

6. Add 2 mL of phenol saturated with extraction buffer. Mix.

7. Centrifuge at 12,000 × *g* for 15–20 minutes at 4°C to separate the phases.

COMMENTS

1. Homogenizing Buffer

0.1 M Tris-HCl, pH 7.5
50 mM KCl or NaCl
10 mM MgCl$_2$

Different tissues or cells might require different buffers and/or lysing procedures. Refer to the literature for specific criteria for lysing the cells of choice.

2. This step pellets nuclei and cell debris.

3. Alternatively, the polysomes can be concentrated by applying the supernatant to a discontinuous sucrose-gradient composed of two layers of 2M and 1.5M sucrose in homogenizing buffer. The gradient is centrifuged at 100,000 × *g* for 3 hours.

4. Extraction Buffer

0.1M Tris-HCl, pH 7.6
0.1M NaCl
1 mM EDTA

5. SDS denatures proteins. It not only inactivates ribonucleases but also disrupts polysome structure thus facilitating the extraction of RNA.

Skin (fingers) is a possible source of RNase.

PROCEDURE	COMMENTS

8. Remove the upper aqueous phase containing the RNA, and repeat the phenol extraction.

9. Extract residual phenol with ether. Remove the ether by aspiration (the ether layer is on top).

10. Evaporate residual ether by passing a stream of air over the surface of the solution.

11. Add 0.2 mL of 3M sodium acetate to a final concentration of 0.3M.

12. Add 2 volumes of cold ethanol and precipitate the RNA at −70°C for 15 minutes or at −20°C for 2 hours.

13. Pellet the precipitated RNA by centrifugation at 12,000 × g for 15−20 minutes at 4°C.

14. Air dry the pellet and dissolve the RNA in sterile application buffer (approximately 500 μL).

14. Application Buffer

 10 mM Tris-HCl (pH 7.6)
 500 mM KCl or NaCl

15. Prepare a 2 mL oligo(dT) cellulose column in a disposable column or Pasteur pipette. Equilibrate the column with application buffer.

15. All equipment and materials (except oligo(dT) cellulose) should be autoclaved before use to denature any contaminating RNases.

16. Load the mRNA sample onto the column.

17. Wash the column with 5× 2 mL of application buffer. Discard the column run off.

17. All of the nonadsorbed materials will be eluted out. The polyA tails of the mRNA bind to the oligo(dT) cellulose at high ionic strength.

18. Elute the column with 4× 2 mL of first elution buffer. Discard the runoff.

18. First Elution Buffer

 10 mM Tris-HCl (pH 7.5)
 100 mM KCl

This step helps to remove nonspecifically bound RNA.

19. Elute the column with 4× 2 mL of second elution buffer. Collect fractions of 2 mL volume.

19. Second Elution Buffer

 10 mM Tris-HCl (pH 7.5)

Bound mRNA is eluted out with low ionic strength buffer.

20. Locate the polyA-tailed mRNA by measuring the optical density of the fractions at 260 nm.

21. Pool the mRNA fractions, and add 3M sodium acetate to a final concentration of 0.3M.

22. Precipitate the polyA-tailed mRNA with 2 volumes of cold ethanol at −70°C for 15 minutes or at −20°C for 2 hours.

23. Pellet the mRNA by centrifugation at 12,000 × g for 15 minutes at 4°C.

24. Dissolve the pelleted RNA in the appropriate buffer.

24. The buffer employed is dependent on the subsequent use to be made of the mRNA.

Oligo(dT) columns may be regenerated by washing with 4× 2 mL of 0.1M KOH, followed by distilled water, until the pH is less than 8. Equilibrate the column with application buffer before reuse.

B. SYNTHESIS OF COMPLEMENTARY DNA (cDNA)

The eukaryotic genome, in comparison to that of prokaryotic cells, is so complex that the direct cloning of specific genes from these cells may be precluded. Instead, the double-stranded DNA molecule to be cloned is synthesized from a mRNA template using the enzyme reverse transcriptase (Efstratiadis and Villa-Komaroff, 1979). Reverse transcriptase is an RNA-dependent DNA polymerase isolated from RNA tumor viruses (Baltimore, 1970; Temin and Mizutani, 1970; Verma, 1977). The classical steps involved in the synthesis of complementary DNA (cDNA) are the following: (1) Using poly(A) mRNA as a template, a 12–18 base oligo(dT) fragment as primer, and the reverse transcriptase from avian myeloblastosis virus (AMV), a single-stranded cDNA copy of the mRNA is synthesized; (2) The mRNA template is removed by alkaline hydrolysis. Single-stranded cDNA can fold back on itself to form a hairpin loop structure, which can act as primer for the synthesis of the second strand. The second cDNA is synthesized using either reverse transcriptase or the Klenow fragment of DNA polymerase I; (3) The single-stranded region of the hairpin structure is removed by digestion with S1 nuclease; (4) The molecule is cloned into the appropriate vector by homopolymeric tailing or with synthetic linkers. Homopolymeric tailing is the method of choice for cloning cDNA, since the probability is high that the cDNA may have the same restriction sites as the synthetic linkers.

The digestion of double-stranded cDNA with S1 nuclease invariably results in the removal of sequences corresponding to the 5'-proximal region of the mRNA (Ulhrich et al., 1977). In order to obtain full-length cDNA, a technique has been developed in which the mRNA·cDNA hybrid is "tailed" prior to the hydrolysis of the mRNA template and the subsequent synthesis of the second cDNA strand (Goddard et al., 1983). Recently, an elegant procedure for the full-length directional cloning of double-stranded cDNA has been described by Okayama and Berg (1982). The following is a generalized procedure for synthesizing double-stranded cDNA.

PROCEDURE   **COMMENTS**

1. To 10 µg (10µL) of polyA-tailed mRNA, add the following:

   5 µL of 500 mM Tris-HCl (pH 8.3)
   5 µL of 500 mM KCl
   5 µL of 80 mM $MgCl_2$
   5 µL of 4 mM dithiothreitol
   5 µL of 1 mg/mL actinomycin D
   5 µL of 10 mM dNTP's
  10 µL of 1 mg/mL oligo(dT)$_{12-18}$

1. Wherever possible, all materials and equipment should be autoclaved to eliminate any contaminating RNases.

Actinomycin D inhibits the synthesis of the second cDNA strand by reverse transcriptase (Verma, 1977). One of the dNTP's should be radioactively labeled so that the progress of the cDNA synthesis can be followed. It is preferable to use ^{3}H-dNTP at this stage, so as not to interfere with the α-^{32}P-dNTP used in the subsequent homopolymeric tailing procedure.

PROCEDURE	COMMENTS

2. Add 40−60 units (2−3 μL) of AMV reverse transcriptase, and incubate for 1−3 hours at 42°C.

3. Stop the reaction by adding 5 μL of 200 mM EDTA and 2% SDS. (Final concentration 20mM EDTA and 0.2% SDS.)

4. Eliminate any unincorporated dNTP's by eluting the sample through a Sephadex G-50 or G-100 mini column (see Section XVII-N, p. 103). Determine the fractions containing the labeled mRNA·cDNA hybrid by scintillation counting of the acid-insoluble material (Section XVII-O, p. 106).

5. Pool the fractions containing the mRNA·cDNA hybrid. Add 3M sodium acetate to a final concentration of 0.3M, followed by the addition of 2 volumes of ethanol. Precipitate the mRNA·cDNA hybrid at −70°C for 10 minutes, or −20°C for 2 hours.

6. Centrifuge at full speed for 10−15 minutes in a microfuge to pellet the hybrid.

7. Add 10−20 dC residues to the 3′ end of the mRNA·cDNA hybrid using the homopolymeric tailing procedure described in Section X, Method 3, p. 53.

8. Resuspend the poly(dC)-tailed hybrid in 25 μL of H_2O, then add 25 μL of 0.6M NaOH (final concentration 0.3M), and incubate at 65°C for 30 minutes.

8. The mRNA template is hydrolyzed at this step.

9. Neutralize the reaction mixture with 1M Tris-HCl (pH 7.0). Check the pH with pH paper.

10. Add 3M sodium acetate to a final concentration of 0.3M, and 2 volumes of ethanol. Precipitate the poly(dC) tailed single-stranded cDNA at −70°C for 10 minutes and pellet by centrifugation at full speed for 10−15 minutes.

11. Purify the poly(dC)-tailed single-stranded cDNA by affinity chromatography on an oligo(dG) column. Pool and precipitate the purified cDNA as in step 10.

11. The step is the same as that described for the isolation of eukaryotic mRNA, (Section X4-A, p. 83) except an oligo(dG) column is used.

12. Resuspend the poly(dC)-tailed single-stranded cDNA in 50 μL of 200 mm Tris-HCl (pH 8.3) containing 200 mM KCl and 80 mM $MgCl_2$.

13. Add 10−30 μg of oligo $(dG)_{12-18}$ to the reaction mixture.

14. Heat to 65°C for 5 minutes and then at 42°C for 30 minutes.

14. This step hybridizes the oligo$(dG)_{12-18}$ to the poly(dC) tail of the cDNA. The oligo-(dG) will then serve as the primer for the synthesis of the second strand.

15. After chilling, add 50 μL of 20 mM DTT with 2 mM of each dNTP.

PROCEDURE	COMMENTS

16. Add 20 units of reverse transcriptase, and incubate for 1–3 hours at 37°C.

16. Klenow fragment (20–50 units) can also be used instead of reverse transcriptase.

17. Digest with S1 nuclease to remove single-stranded ends.

17. Add 2000 units/mL of S1 nuclease to the reaction mixture in the presence of 0.5% SDS and incubate for 20 minutes at 37°C.

18. Phenol/ether extract the reaction mixture (Section I).

19. Precipitate the blunt-ended double-stranded cDNA as described in step 10.

20. Clone the blunt-ended cDNA into a cloning vector using homopolymeric tailing as described in Section X, Method 3, p. 53.

C. PREPARATION OF DENATURING GELS

The electrophoretic behavior of nucleic acids, especially RNA and single-stranded DNA, is influenced by their size, their charge, and also by their conformation (secondary structure). For the accurate determination of the molecular weight of single-stranded nucleic acid, it is necessary to eliminate the differences caused by conformation. This is achieved by electrophoresis in denaturing gels. These gels are prepared by the incorporation of such denaturing agents as sodium hydroxide, urea, formamide, methyl mercuric hydroxide, formaldehyde, and sodium dodecyl sulphate (Studier, 1973; Lehrach *et al.*, 1977). For agarose gels, 30 mM sodium hydroxide, 5 mM mercuric hydroxide or 2.2M formaldehyde are often used. Formamide is not used in agarose gels because it dissolves agarose. Since methyl mercuric hydroxide is highly toxic, extra precautions must be taken while handling this chemical. 7M urea, 98% formamide and 0.1% SDS are used with polyacrylamide gels.

METHOD 1: METHYL MERCURIC HYDROXIDE AGAROSE GELS

PROCEDURE	COMMENTS

1. Suspend 0.5 gm of agarose in 50 mL of 1× electrophoresis buffer (Tris-borate or Tris-acetate).

1. This will give a final concentration of 1% agarose in the gel. Use same buffer formulations as recommended for DNA separations (Section V-A, Method 1, p. 18).

2. Melt the agarose by bringing to a boil either in a microwave oven or on a hot plate.

3. Cool the melted agarose solution to ≃ 60°C.

4. Add 250 μL of 1M methyl mercuric hydroxide solution to the molten agarose and mix.

4. Methyl mercuric hydroxide is commercially obtained as a 1M solution. The final concentration of methyl mercuric hydroxide in the gel is 5 mM.

CAUTION: *Methyl mercuric hydroxide is very toxic.*

5. Cast the gel as described in Section V, p. 18.

METHOD 2: FORMAMIDE POLYACRYLAMIDE GELS

<table>
<tr><td>**PROCEDURE**</td><td>**COMMENTS**</td></tr>
</table>

1. Prepare polyacrylamide gels of different concentration which contain 98% formamide as follows (Table 1.7):

TABLE 1.7 Preparation of Formamide-Polyacrylamide Gels

Stock Solution (in mL)	Concentration of Gel			
	5%	8%	12%	20%
B	6.7	10.7	16.0	26.7
A	32.9	28.9	23.6	12.9
C	0.4	0.4	0.4	0.4

Final volume is 40 mL.

2. Add 60 μL of TEMED. Mix.

3. Cast and prepare the gel as described in Section V, p. 18.

4. Load the sample into the gel and run the gel using 20 mM NaCl in water as electrophoresis buffer.

1. Preparation of Stock Solutions

A Dissolve 3.68 gm of diethylbarbituric acid in 1 litre of deionized formamide. Adjust to pH 9.0 with HCl.

B 30% acrylamide. Dissolve 28.6 gm of acrylamide and 1.4 gm of Bis-acrylamide in 100 mL of solution A.

C 18% ammonium persulfate (freshly prepared).

2. Deaeration is not required.

4. The sample should be suspended in solution A containing 0.1% bromophenol blue and 20% sucrose. Circulation of the electrophoresis buffer between chambers is required.

METHOD 3: 7M UREA POLYACRYLAMIDE GEL

<table>
<tr><td>**PROCEDURE**</td><td>**COMMENTS**</td></tr>
</table>

1. Different concentrations of polyacrylamide gel containing 7M urea are prepared, using the following formulae (Table 1.8):

TABLE 1.8 Preparation of Formamide-Polyacrylamide Gels

Stock Solution	Concentration of Gel			
	5%	8%	12%	20%
Urea (gm)	16.8	16.8	16.8	16.8
A (mL)	6.7	10.7	16.0	26.7
B (mL)	0.8	0.8	0.8	0.8
C (mL)	4.0	4.0	4.0	4.0

Make up to 40 mL with double distilled water.

2. Deaerate the solution under vacuum.

3. Add 15 μL of TEMED to the deaerated polyacrylamide solution. Mix.

4. Cast and assemble the gel as described in Section V, p. 18.

1. Preparation of Stock Solutions

A 30% acrylamide (28.6 gm of acrylamide, 1.4 gm of Bis-acrylamide).

B 3% ammonium persulfate (freshly prepared).

C 10× concentrated (Tris-borate) electrophoresis buffer.

2. The solution is deaerated when bubbling stops.

5.   Load the gel with the sample.

5.   Suspend the sample in $1\times$ electrophoresis buffer containing 7M urea, 0.1% marker dye, and 20% sucrose.

6.   Run the gel under the same conditions as polyacrylamide gels.

6. $1\times$ electrophoresis buffer is the same buffer used to prepare the gel.

XVI. DNA Sequencing Methods

Two basic methods are used to determine the nucelotide sequence of a defined DNA fragment. Frederick Sanger and his colleagues (1977) developed the dideoxy chain termination DNA sequencing method, while independently, Allan Maxam and Walter Gilbert (1977) developed the "chemical" method. Both methods require the use of small, defined fragments of DNA that are obtained by restriction endonuclease digestion.

Of the two procedures, the Sanger method of chain termination sequencing is considered to be more simple, rapid, and accurate (Davies, 1982). The disadvantage of the dideoxy procedure is that the DNA to be sequenced must be in a single-stranded form, and that for every 200–300 bases of sequence, a specific primer, complementary to the 3'-end must be prepared. Several strategies have been developed to overcome this problem (Guo and Wu, 1983). Currently the most popular one is to clone the DNA fragment to be sequenced into an M13 (or fd) cloning vector, which will produce a single-stranded DNA as template. The "universal" primer used in this system is commercially available. The chemical method is considered to be easier for beginners, as neither template DNA nor primer DNA must be prepared, and labile enzymes are not involved. However, the method is time-consuming, it requires more starting materials (i.e., DNA and labeled nucleotides), and the reagents must be pure and fresh.

The following gives a brief description and background information pertaining to the two techniques, so as to familiarize the reader with their use. Exhaustive reviews have been completed by others (Davies, 1982; Hindley, 1983), and a recent article by Deininger (1983) gives excellent tips on DNA sequencing. For complete details of the Maxam and Gilbert method, the reader should refer to the excellent article "Sequencing End-labeled DNA with Base-Specific Chemical Cleavage" by Allan M. Maxam and Walter Gilbert in *Methods in Enzymology*, Vol. 65, 1980. Other chapters (e.g., Chapter 4) give specific examples for the use of the dideoxy chain termination method.

A. DIDEOXY CHAIN TERMINATOR METHOD

The dideoxy chain termination method is based upon the ability of DNA polymerase to synthesize a radioactively-labeled complementary copy of a single-stranded (ss) DNA template, using a short complementary ssDNA fragment as primer, and to incorporate randomly a dideoxynucleotide (ddNTP) instead of a deoxynucleotide (dNTP). Since dideoxynucleotides lack the 3'-hydroxyl group necessary for the formation of the next phosphodiester bond during chain elongation, chain elongation is therefore terminated whenever a dideoxynucleotide is incorporated into the strand. That is, if both dideoxynucleotide (e.g., ddATP) and the corresponding deoxynucleotide (e.g., dATP) are present, chain elongation will terminate randomly according to the site or position at which the ddNTP is incorporated. The large fragment of DNA polymerase I (Klenow fragment), which lacks $5' \rightarrow 3'$ exonuclease activity, is generally used in this reaction to keep the 5'-end of the primer intact.

For sequence analysis, four separate reactions (one for each nucleotide) are carried out simultaneously. Each reaction mixture contains the Klenow fragment, all four deoxynucleotides, one or more of which is labeled with ^{32}P and only *one* of the four dideoxynucleotides. The concentration of the dideoxynucleotide in the reaction mixture is adjusted so that the incorporation of the dideoxynucleotide (therefore chain growth termination) occurs randomly at each possible site. Thus, in each reaction mixture, a population of partially synthesized, radioactively-labeled DNA molecules having the same 5'-end (i.e., primer) and the same base at its 3'-end, but varying in length, can be found. By varying the ratio between corresponding dideoxynucleotide and deoxynucleotide (i.e., ddATP:dATP) different results are achieved; a high ratio will intensify bands in a region near the primer but more distal information will be lost. A low ratio will result in longer sequences, but the resolving capacity of the sequencing gel becomes a limiting factor.

PROCEDURE

1. Primer/Template Annealing Reaction

To a 500 μL microfuge tube add the following:

5 μL (0.5 μg) single-stranded template DNA
2 μL (5 μg) primer DNA fragment
1 μL of concentrated Klenow fragment reaction buffer
4.5 μL of double distilled H_2O

Pulse spin the microfuge tube (3−4 seconds), so the components will be mixed at the bottom of the tube.

COMMENTS

1. Concentrated Klenow Fragment Reaction Buffer

70 mM Tris-HCl, pH 7.6
70 mM MgCl
500 mM NaCl

Two procedures for preparing single-stranded DNA template are described in Section IIIA, Chapter 4, p. 169. The bacteriophage M13 method is most commonly used.

A primer DNA fragment is a DNA fragment with a sequence complementary to a region of the template DNA. After annealing to that region, it provides a 3'-hydroxyl group for DNA chain elongation using the Klenow fragment of polymerase I. The "universal" primer of the M13 sequencing system is a synthetic DNA fragment of 15−17 bases, which is complementary to a region of M13 DNA adjacent to the 3'-end of the multiple cloning sites (MCS). Thus, when chain elongation is initiated with the "universal" primer, it will pass through the MCS region, incorporating the sequence of any DNA fragment inserted at this position. The synthetic "universal" primer for M13 sequencing system is commercially available.

2. Cap the tube and incubate in a 85°−90°C waterbath for 5 minutes.

2. This can also be done at 55°−60°C for 1−2 hours.

3. Leave the microfuge tube containing the primer/template mixture at room temperature for 45 minutes.

3. Slow equilibration to room temperature will ensure proper annealing. If the annealed primer/template mixture is not used immediately, it can be stored at −20°C.

4. While the primer/template is annealing, prepare the following four 500 μL microfuge tubes. Add the components, without mixing to each tube.

Tube A: 1 μL A mix and
       1 μL ddATP (1 mM)
Tube C: 1 μL C mix and
       1 μL ddCTP (0.35 mM)
Tube G: 1 μL G mix and
       1 μL ddGTP (0.7 mM)
Tube T: 1 μL T mix and
       1 μL ddTTP (2 mM)

4. Add each of the two components down opposite walls of the microfuge tube. Surface tension will keep them from mixing.

If labeling with [α^{32}P]-dATP, use the following mixtures:

Ingredient Added	Volume Added to Mix			
	A	C	G	T
0.5 mM dCTP	20 μL	1 μL	20 μL	20 μL
0.5 mM dGTP	20 μL	20 μL	1 μL	20 μL
0.5 mM dTTP	20 μL	20 μL	20 μL	1 μL
Concentrated Klenow reaction buffer	20 μL	20 μL	20 μL	20 μL

Alternatively, if labeling with [α^{32}P]-dCTP, the following mixtures are used:

Ingredient Added	Volume Added to Mix			
	A	C	G	T
0.5 mM dATP	1 μL	20 μL	20 μL	20 μL
0.5 mM dGTP	20 μL	20 μL	1 μL	20 μL
0.5 mM dTTP	20 μL	20 μL	20 μL	1 μL
Concentrated Klenow reaction buffer	20 μL	20 μL	20 μL	20 μL

5. To the microfuge tube containing the annealed primer/template (step 3), add the following:

1 μL (10 μCi)[α^{32}P]-dATP (or [α^{32}P]-dCTP)
1 μL 0.1M dithiothreitol
1 μL Klenow fragment (1 unit/μL)

5. If the [α^{32}P]-dATP or [α^{32}P]-dCTP is in ethanol, dry down by evaporation and resuspend in H$_2$O before use.

6. Mix the tube contents by passing them in and out of a micropipetter tip.

7. Dispense 3 μL of the mixture to each of the four tubes prepared in step 4. Pulse-spin to mix all the components at the bottom of the tube.

8. Incubate at room temperature for 15 minutes.

9. Add 1 μL of 0.5 mM dATP (or dCTP) to each tube and incubate the reaction mixture for another 15 minutes at room temperature.

9. This serves to extend chains which were terminated prematurely because of low dATP (dCTP) concentration.

10. STOP the reaction by adding 10 μL of formamide-dye mix.

10. Formamide-dye Mix

   0.1 g xylene cyanol FF
   0.1 g bromophenol blue
   0.75 g Na$_2$EDTA·2H$_2$O (20 mM)
   100 mL deionized formamide

11. Denature DNA in the reaction mixtures by heating in boiling water for at least 3 minutes.

12. Immediately load onto a sequencing polyacryla-mide gel.

12. Polyacrylamide-sequencing Gels

a. An 8% polyacrylamide-7M urea gel (see Section XV, Method 3, p. 89), having either 0.4 mm × 40 or 80 cm × 20 cm dimensions, is commonly used. Other concentrations of polyacrylamide have also been used. Lower concentration gels give wider band spacing and are often used for long sequencing runs. They have the disadvantage of being fragile and hard to handle. Higher concentration gels give narrow band spacing. Gels with concentrations up to 20% polyacrylamide have been used when sequencing the first 25 nucleotides of a DNA fragment using the Maxam and Gilbert method.

b. The running buffer is 90 mM Tris-borate (pH 8.3) plus 1 mM EDTA.

c. Since the sequencing gel is so thin (0.4 mm), loading the sample onto the gel might pose some difficulty to some beginners. It might be easier using a device shown in Figure 1.13.

d. If the bases to be sequenced number less than 100, load the entire sample onto the gel and electro-phorese until the orange tracking dye, xylene cyanol, has migrated approximately one-third to one-half the

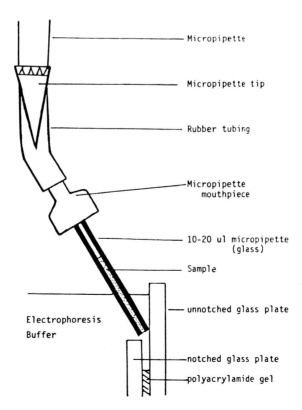

Micropipette

Micropipette tip

Rubber tubing

Micropipette mouthpiece

10-20 ul micropipette (glass)

Sample

unnotched glass plate

Electrophoresis Buffer

notched glass plate

polyacrylamide gel

FIGURE 1.13 Sample applicator used in sequencing gels.

distance of the gel. Xylene cyanol migrates with fragments 70 bases long on an 8% polyacrylamide gel. If 200 or more bases are to be sequenced, then the sample should be divided into 3 portions. One portion should be run on a 20% polyacrylamide gel in order to sequence the first 30 bases. The second portion, which is used to identify base numbers 25–100, should be loaded onto a 8% polyacrylamide gel and electrophoresed until the tracking dye, bromophenol blue, reaches the bottom. Load the third portion of the sample, used to identify the last 100–150 bases, onto a second 8% polyacrylamide gel and electrophorese. Note the time required for the xylene cyanol to run to the bottom of the gel (the bromophenol blue dye will have run off), and continue to electrophorese further for one-half of this time. Alternatively, portions 2 and 3 of the sample may be run on the same 8% polyacrylamide gel. Load portion 3 onto the gel and run until the xylene cyanol reaches the bottom of the gel. Then load portion 2 in an unused slot, and continue to electrophorese the bromophenol blue marker until this portion reaches the bottom.

13. Electrophorese the samples using 30 mA of current ($\approx$1.3–1.5 kv).

13. The gel will rapidly warm to 55°–60°C. This high temperature is essential to achieve complete denaturation of DNA sequences (in complement with the urea in the gels).

NOTE: *It may be advantageous to sandwich the gel-forming plates between two aluminum plates (2 mm thick) of the same size. This ensures that the heat is evenly distributed throughout the gel and prevents possible cracking of the glass plates.*

14. After electrophoresis, the notched plate is gently pried off with a thin spatula, leaving the gel on the unnotched plate.

15. Fix the DNA by treating the gel with 10% acetic acid for 5 minutes and then wash with deionized water.

16. Carefully dry the gel (still on the plate) with blotting paper. Be careful not to "stick" the gel onto the blotting paper.

17. Cover the gel with Saran Wrap and proceed with autoradiographic procedures (Section XIII, p. 79).

B. CHEMICAL (MAXAM AND GILBERT) METHOD

With this method, the DNA fragment to be sequenced is purified and labeled at one end only. This is achieved by first labeling both ends of the DNA fragment (see

Complete the following steps simultaneously with all four reaction mixtures.

	REACTION MIXTURE NUMBER			
	Number 1 (C)	Number 2 (T+C)	Number 3 (G)	Number 4 (G+A)
1. Add	15 µL 5M NaCl 1 µL CT DNA 5 µL ^{32}P-DNA	10 µL deionized water (dH$_2$O) 1 µL CT DNA 10 µL ^{32}P-DNA	200 µL DMS buffer 1 µL CT DNA 5 µL ^{32}P-DNA	10 µL dH$_2$O 1 µL CT DNA 10 µL ^{32}P-DNA
2.	Chill to 0°C.			
3. Add	30 µL hydrazine	30 µL hydrazine	1 µL DMS	2 µL pyridinium formate
4.	Incubate for 10 minutes (±5) at 20°C. (To obtain short fragments incubate for 15 minutes, long fragments for 5 minutes.)			20°C, 60 ± 20 min
5. Add	200 µL HZ stop 750 µL ethanol	200 µL HZ stop 750 µL ethanol	500 µL DMS stop 750 µL ethanol	Freeze Lyophilize
6.	Chill on dry ice for 5 minutes			Add 20 µL dH$_2$O and lyophilize again.
7.	Centrifuge in a microfuge at maximum speed for 5−10 minutes			
8.	Decant ethanol			
9. Add to pellet:	250 µL 0.3M chilled sodium acetate 750 µL chilled ethanol			
10.	Repeat steps 6 to 8.			
11.	Wash pellet with 80% cold ethanol and repellet			
12.	Dry pellet under vacuum			
13.	Add 100 µL of 1.0 M piperidine			
14.	Cap the microfuge tube and incubate at 90°C for 30 minutes; (Put weight on the caps so they won't pop)			
15.	Lyophilize sample			
16.	Add 10 µL deionized water and lyophilize			
17.	Repeat step 16			
18.	Add 10 µL formamide-NaOH dyes			
19.	Incubate at 90°C for 1 minute			
20.	Quickly chill to 0°C			
21.	Load polyacrylamide gel and run as described in dideoxy chain terminator method, p. 93.			

DMS BUFFER: DIMETHYL SULFATE BUFFER

0.05M sodium cacodylate
0.01M magnesium chloride
0.001M EDTA
Adjust to pH 8.0 with NaOH

DMS STOP: DIMETHYL SULFATE STOP

1.5M sodium acetate (pH 7.0)
1.0M mercaptoethanol
100 µg/mL tRNA

HZ STOP: HYDRAZINE STOP

0.3M sodium acetate (pH 5.5)
0.1mM EDTA
100 µg/ml tRNA

FORMAMIDE-NaOH-DYES

Deionized formamide: 100 mL of formamide with 5 g mixed-bed, ion-exchange resin (e.g., Bio-Rad AG501-X8, 20−50 mesh) for 39 minutes. Remove resin by filtration.
10mM NaOH
 1mM EDTA
0.1% (w/v) xylene cyanol
0.1% (w/v) bromophenol blue

end-labeling, Section XII, this chapter), then either (a) denaturing the DNA fragment and separating the two single strands on a polyacrylamide gel, or (b) cutting the DNA fragment with a restriction enzyme and then purifying the products electrophoretically. The purified DNA fragment, labeled at a single end is then subjected to a set of at least four separate chemical treatments, the end result of which is cleavage of the DNA backbone at one, or in some cases two, of the four bases. Each procedure involves three steps: (a) modification of a base, (b) removal of the modified base from its sugar, and (c) DNA strand scission at that sugar. If complete chemical modification of a particular base is achieved, only one short labeled fragment is obtained. However, if conditions are chosen such that only partial modification is possible (i.e., one out of all possible bases on the same DNA molecule is modified), a family of DNA molecules of different lengths with the same labeled ends is obtained. The extent of chemical modification can be controlled by varying the time of the reaction. Short reaction times give low levels of modification, and allow analysis of long sequences, while long reaction times will lead to high levels of modification and thus, short fragments. These molecules are separated and analyzed according to size on a sequencing polyacrylamide gel. By running a set of samples obtained with the four different series of chemical reactions in parallel on the same gel it is possible to determine the nucleotide sequence of the DNA fragment. A synopsis of the procedure is given below.

XVII. Miscellaneous Information and Methods

A. UNITS OF MEASUREMENT

$1 \text{ mL} = 10^{-3} \text{ L}$

$1 \text{ }\mu\text{L} = 10^{-6} \text{ L}$
$\quad\quad = 10^{-3} \text{ mL}$

$1 \text{ kilogram (kg)} = 1000 \ (10^{3}) \text{ gram (g)}$
$1 \text{ milligram (mg)} = 0.001 \ (10^{-3}) \text{ gram (g)}$

$1 \text{ microgram }(\mu\text{g}) = 10^{-6} \text{ g}$
$\quad\quad\quad\quad\quad\quad = 10^{-3} \text{ mg}$

$1 \text{ nanogram (ng)} = 10^{-9} \text{ g}$
$\quad\quad\quad\quad\quad\quad = 10^{-6} \text{ mg}$
$\quad\quad\quad\quad\quad\quad = 10^{-3} \text{ }\mu\text{g}$

$1 \text{ picogram (pg)} = 10^{-12} \text{ g}$
$\quad\quad\quad\quad\quad\quad = 10^{-9} \text{ mg}$
$\quad\quad\quad\quad\quad\quad = 10^{-6} \text{ }\mu\text{g}$
$\quad\quad\quad\quad\quad\quad = 10^{-3} \text{ ng}$

kb $=$ kilobase
1 kb $=$ 1000 base pairs (bp) of nucleic acid
1 kb $=$ 0.66 megadaltons (Md), double-stranded DNA (1 bp $=$ 660 daltons)
1 kb $=$ 0.34 Md, single-stranded RNA (1 bp $=$ 340 daltons)

$1 \text{ Md} = 10^{6} \text{ daltons}$

$1 \text{ dalton} = 1.65 \times 10^{-24} \text{ g}$

B. WEIGHTS

MOLECULAR WEIGHT (MW)	is the sum of the atomic weights of the constituent atoms in a compound;

e.g., molecular weight of NaH_2PO_4

$$MW = 23+(1\times2)+31+(16\times4)$$
$$= 120$$

GRAM-MOLECULAR WEIGHT	is the weight in grams of a compound numerically equal to its molecular weight;

e.g., one gram-molecular weight (1 mole) of NaH_2PO_4 = 120 grams

GRAM-EQUIVALENT WEIGHT	is the weight in grams of a compound that will produce or react with one mole of hydrogen;

e.g., 1 gram-equivalent of
$$NaH_2PO_4 = 120/2$$
$$= 60$$

C. CONCENTRATIONS

Concentrations of solutions are often expressed as weight per weight (w/w), weight per volume (w/v), volume per volume (v/v).

1% (w/w) (PERCENT BY WEIGHT)	solution contains 1 g solute per 100 g solution (i.e., 1 g solute + 99 g solvent).
1% (w/v) (PERCENT WEIGHT PER VOLUME)	solution contains 1 g solute per 100 mL of solution.
1% (v/v) (PERCENT BY VOLUME)	solution contains 1 mL of solute (liquid) per 100 mL of solution.
1 molar (1M)	solution contains 1 gram-molecular weight per 1000 mL (1 L) of solution.
1 normal (1N)	solution contains 1 gram-equivalent weight per 1000 mL (1 L) of solution (with NaOH, HCl, N = M).

In preparing a solution by volume, it is advisable to first dissolve the proper weight of solute in a small volume of solvent. Then, after the solute has completely dissolved, adjust the volume of the solution to the desired amount.

D. PREPARATION OF A DILUTE SOLUTION

Since mistakes in weighing out microquantities often introduce significant sources of experimental error, the dilution of a more concentrated solution is a simpler and more accurate alternative. For example, a solution of 1 μg (microgram) per mL may be prepared by weighing out 1.0 mg of the solute and dissolving it in 1000 mL of solvent. If a concentration of 1 ng (nanogram) per mL is required, the above solution may then be diluted by 10^3.

E. PREPARATION OF BUFFERS (e.g., TRIS)

A buffer is an aqueous solution containing a mixture of a weak acid and a weak base (usually the corresponding conjugate of the acid). It is able to resist ("buffer") the effects of externally supplied H^+ and OH^- groups on its pH. Therefore, buffers are used primarily for pH maintenance. The buffering capacity of a particular mixture is maximal at nearly equal concentrations of its component acid and base. It increases (or decreases) linearly with increases (or decreases) in their concentration.

The compound Tris (hydroxymethyl) aminomethane (Tris) is an ingredient common to most of the buffers used in this manual. Its popularity stems from its high buffering capacity between pH 7 and pH 9, a range ideal for most nucleic acid work. In addition, it shows little interaction with metal ions that may serve as enzyme co-factors.

PROCEDURE

1. Place sufficient distilled H_2O (dH_2O) in an appropriately sized beaker to give approximately 75% of the final volume of buffer.

2. Place the beaker containing the stirring bar on a magnetic stirrer. Insert a pH meter probe.

3. Add each buffer component sequentially while stirring.

4. After all of the components are dissolved, add sufficient dH_2O to give 90% of the final volume. Continue mixing for 5 minutes.

5. Determine the pH of the solution.

If lower than the desired value adjust it upwards by the drop-wise addition of an alkali solution; if higher use an acid solution.

NOTE: *Continue stirring throughout this step.*

6. Mix for an additional 5 minutes. When the pH has stabilized, adjust the solution to its final volume with dH_2O.

COMMENTS

1. The volume may increase as each component is dissolved, particularly when liter quantities are being prepared.

2. The pH reading of the meter should be corrected with standard buffer having a pH near that of the new buffer prior to use.

Since proton (H^+) activity is thermosensitive, the pH of a solution will vary with its temperature. Thus the buffer solution should be prepared at the same temperature at which it will be used.

4. This adjusts the solution to its approximate final molarity.

5. For most buffers, solutions of NaOH and HCl or acetic acid are adequate for these adjustments. Their concentrations are determined by the molarity of the buffer: if high, use high concentration ("strong") acid and base solutions.

6. The pH of the buffer is not generally significantly altered by this addition.

F. PREPARATION OF SATURATED PHENOL SOLUTIONS

Phenol is an organic acid that is frequently used to selectively remove proteins (Rushizky *et al.*, 1963) and denatured DNA (Currier and Nester, 1976) from aqueous solutions. In particular, it can be used to "clean-up" cell lysates, and at the same time to enrich for covalently-closed duplex (plasmid) DNA. To reduce its solubility in the aqueous phase and to effect its neutralization, phenol is usually saturated with the major component of the aqueous solution (e.g., TE buffer, 3% NaCl).

Many authors recommend the use of redistilled ("ultra pure") phenol, especially if the DNA is to be subjected to subsequent restriction endonuclease analysis. In this instance, the collection of the fraction coming off the still at 160°C will assure that the phenol is free of heavy metal contaminants and inhibitors (e.g., H_3PO_2), which may alter the integrity of nucleic acids through degradation or cross-linking. Other workers routinely add the antioxidant 8-hydroxyquinoline (final concentration 0.1%) to their phenol preparations. This chemical is also a weak metal ion chelator (Kirby, 1968). In our experience however, the use of crystalline analytical reagent grade phenol (mp 39.5°−41°C) without the antioxidant, yields adequate substrates for most endonucleases.

PROCEDURE

COMMENTS

CAUTION: *Phenol is poisonous and caustic. Always wear gloves when handling phenol or its solutions. If skin contact is made rinse under water until the whiteness disappears.*

1. Melt crystalline phenol at 50−60°C.

1. It is desirable to store crystalline phenol at −20°C to inhibit its oxidation.

2. Check the color of the liquified phenol. If very pink or brown, discard and replace with material from a fresh bottle.

2. Highly-colored phenol solutions give poor extractions.

3. Add an equal volume of saturating solution (e.g., TE buffer, 3% NaCl), and mix vigorously.

3. Do this before crystals begin to form.

Some suppliers recommend that phenol be saturated with water, instead of Tris buffer, since such buffers may cause the rapid formation of breakdown products such as aldehydes, esters, anhydrides.

4. Let stand at room temperature. If two layers are not formed within 30 minutes add more saturating solution and mix.

5. Once saturated, pour both layers into a glass separatory funnel and leave at room temperature overnight.

6. Run off the lower phenol layer into a brown glass bottle for storage. Take a small amount of the aqueous phase as well to "cap" the phenol, thereby inhibiting oxidation.

6. Alternatively, the saturating (aqueous) solution and liquified phenol may be mixed directly in a brown bottle. In this instance be certain to use the *lower* layer for extraction.

If placed in a screw-cap bottle, such preparations may be stored at room temperature for 2−3 weeks.

7. Check the solution periodically for color changes.

G. PREPARATION AND USE OF RNase SOLUTIONS

Two types of ribonuclease (RNase) are used to "remove" RNA from nucleic acid preparations: ribonuclease A from bovine pancreas and ribonuclease T1 from *Aspergillus oryzae*. By tradition, RNase A has been the enzyme of choice, but because RNase T1 is believed to be more efficient in digesting large molecular weight RNAs, recent procedures have used a combination of the two (Birnboim, 1983) or RNase T1 alone (Thompson *et al.*, 1983). A possible disadvantage to the use of RNases is suggested by the observation of Lehman *et al.* (1962) that the activity of certain deoxyribonucleases (DNases) is inhibited by transfer-RNA. Elimination of t-RNA may therefore result in some degradation of DNA through the "activation" of these enzymes.

Since commercially available RNases are frequently contaminated with DNase activities, RNase solutions are usually heated prior to their use to destroy the DNases. It is claimed that RNase A contains a residual DNase activity that cannot be completely eliminated by heating (NACS™ Applications Manual, Bethesda Research Labs), further supporting the use of RNase T1.

RNase A: dissolve to 2–5 mg/mL in distilled water, or 5 mM Tris-HCl (pH 8.0), or 0.15M sodium chloride.

RNase T1: dilute to 50 units/mL in distilled water or 5 mM Tris-HCl (pH 8.0).

Place in boiling water (100°C) for 10 minutes and cool. Add the solution to the preparations to give a final concentration of 0.4–3.5 μg/mL RNase A or 5 units/mL of RNase T1. Incubate at 37°C for 15 to 60 minutes (longer times required for RNase A).

H. SILICONIZATION PROCEDURE FOR LABORATORY GLASS AND PLASTICWARE

Equipment is siliconized before use to minimize the loss of DNA through adsorption (e.g., glasswool) or sticking.

PROCEDURE

1. Prepare a 5% (v/v) solution of dichlorodimethylsilane in one of the following solvents: chloroform, toluene, carbon tetrachloride or methylene chloride.

2. For treating test tubes or microfuge tubes, fill the container with the above solution for 1–2 minutes. For treating glasswool or glass rods, soak the item in the solution for 1–2 minutes.

3. Drain the container and save the solution. Invert the container and air dry. For glasswool, squeeze the solution out and air dry.

4. Rinse the siliconized materials with deionized water a few times, and air dry.

COMMENTS

CAUTION: *Dichlorodimethylsilane is toxic. Wear gloves and perform all steps in a fume hood.*

Dichlorodimethylsilane can be obtained from Fisher Scientific and Eastman (Kodak).

I. TESTING THE GENOTYPE OF BACTERIAL STRAINS

Most bacteria used in molecular biology carry a number of phenotypic markers that serve as a means of (a) assessing the purity of a particular culture; (b) detecting the presence of a particular extrachromosomal particle (plasmid, cosmid, phage); and (c) selecting a particular recombinant molecule. Chief among these markers are antibiotic resistance markers, resistance to specific chemicals or agents (uv light, colicins, etc.) and auxotrophic requirements.

The presence of the genes determining these properties is tested in various ways. Resistance (or sensitivity) to antibiotics may be easily determined using the disc diffusion or agar dilution assays. A synopsis of these techniques is described by Lorian (1980).

The auxotrophic ("nutritional") requirements of a strain are confirmed by spreading it on nutritionally-defined media. Comparison of the growth achieved on the basic medium, relative to that on medium supplemented with the appropriate sugars, amino acids, vitamins and nucleotide precursors, enables determination of the strain's genotype. The minimal medium devised by Davis and Mingioli (1950) serves well as a basal medium for these tests.

Various other markers (e.g., colicin immunity, uv sensitivity, recombination efficiency) are tested, using a variety of other procedures. Many of these procedures are elegantly described by Miller (1972).

J. STORAGE OF STRAINS AND PHAGE

Strains used on a daily basis may be stored on slants, plates or stabs of an appropriate solid medium (e.g., nutrient agar). Screw-capped or wax-sealed tubes should be used for the slants. Petri plates may be sealed with parafilm to prevent their dehydration. Many *Enterobacteriaceae* may be stored at room temperature for several months using this procedure, although storage at 4°C, away from drying influences such as circulating fans, prolongs the shelf life of the strains.

Enterobacteria may be stored in small aliquots at $-70°C$ (add 40% glycerol to a growth medium) for many years. Prolonged storage (upwards to five years for *E. coli*) may also be achieved by soaking sterile 4×0.6 cm strips of Whatman No. 1 paper with heavy suspensions of the bacterium in 40% glycerol containing 3% trisodium citrate. The strips are sealed in sterile polyethylene bags then "filed" at $-70°C$ until needed. For convenience, the soaking may be completed after placing the strips in the plastic bag (use 8−10 drops of a 1 mL suspension per strip). To recover the strain, the paper strip is simply streaked over the surface of an enriched agar plate or immersed in sterile broth. This procedure facilitates the storage of large numbers of cultures in a small space.

For other species (e.g., *N. gonorrhoeae*) storage at $-70°C$ or lyophilization is the only effective way to maintain strains for long periods (use a sterile solution of 5% glutamine and 5% BSA for storage at $-70°C$). Provided that initial cell concentrations are high, frozen stock cultures may be thawed, a small aliquot removed with a sterile swab, and refrozen without a significant loss of cells.

Lyophilization is the method of choice for long term storage for most strains. Phage may be stored in saline or broth over one drop of chloroform for prolonged periods at 4°C.

K. SOURCES OF CLONING VECTORS AND HOST STRAINS

Some of the more commonly used collections are listed below.

CLONING VECTORS, PLASMIDS AND HOST STRAINS	Recombinant DNA Host Vector Collection American Type Culture Collection 12301 Parklawn Drive Rockville, Maryland 20852
E. COLI STRAINS	B.J. Bachmann *E. coli* Genetic Stock Center Department of Microbiology Yale University School of Medicine 310 Cedar Street New Haven, Connecticut 06510
BACILLUS STRAINS, PLASMIDS, AND CLONING VECTORS	Daniel Ellis, Curator *Bacillus* Genetic Stock Center The Ohio State University Department of Microbiology 484 West 12th Avenue Columbus, Ohio 43210

Many cloning vectors and host strains may be purchased from commercial sources.

L. STORING DNA

A variety of agents, such as intercalating dyes, heavy metals, phenol, ether, and nucleases, may cause damage to DNA upon storage. Therefore, the DNA preparation should be as "clean" as possible, especially for long-term storage. Storage in TE buffer, pH 8.0, at 4°C, has proved a good method for routine storage (good implies a minimum of nicking). DNA may be stored at −70°C for long periods. It should be noted that freezing and thawing DNA is an excellent way to introduce nicks, so this should be avoided. (Storage at −20°C is not recommended. More nicks may be introduced at this temperature since many −20°C freezers are "frost-free" (i.e., have frost-defrost cycles.)

M. DETERMINING DNA CONCENTRATION

Several methods are used for determining the concentration of DNA in a solution. As a rule of thumb, double-stranded nucleic acids have a concentration of 50 μg/mL if the optical density at 260 nm is 1.0. For a more exact quantitation of concentration, the ratio of the optical density readings at 260 and 280 nm is determined. Pure DNA preparations should have an OD_{260}/OD_{280} of 1.8. Ratios significantly lower indicate protein or phenol contamination, while a higher ratio may indicate RNA contamination.

Oftentimes however, DNA is present in quantities too small to measure. In these instances, the DNA concentration is estimated by comparing the ethidium bromide-mediated fluorescence of the unknown sample with that of samples of known DNA concentrations under shortwave uv light.

PROCEDURE	COMMENTS
1. Prepare a 1% agarose plate supplemented with a final concentration of 0.5 μg/mL of ethidium bromide.	**1.** Ethidium bromide-saturated filter paper may be used instead of agarose. Alternatively, the DNA solutions may be directly diluted into an ethidium bromide solution, spotted onto Saran Wrap, and the fluorescence compared (Schleif and Wensink, 1981).
2. Spot the plate with equal volumes of the unknown sample and of dilutions (1–20 μg/mL) of the reference solution.	**2.** Calf thymus DNA or salmon sperm DNA may be used as the reference standard.
3. Allow the spots to soak in for 15 minutes (or longer if the preparation is not pure).	**3.** A longer drying period will allow contaminating molecules to diffuse into the agarose away from the DNA. These molecules include those that quench fluorescence (e.g., SDS, sucrose) or those that cause increased fluorescence (e.g., RNA).
4. Visualize the DNA under short-wave uv (see Section VII, p. 29). Compare the intensity of the reference DNA with the sample DNA to estimate the concentration.	**4.** It should be remembered that contaminating RNA or single-stranded DNA enhances fluorescence.

N. PREPARATION OF GEL FILTRATION COLUMNS

Gel filtration, also known as molecular-exclusion or molecular sieve chromatography, is a technique used for separating a heterogeneous mixture of molecules based on their size differences (Porath and Flodin, 1959). Porous beads (gel particles), such as Sephadex (G series), Bio-gel (P series) or agarose (Bio-gel A series), are often used for such a purpose. The pore size within the beads is regulated by varying either the degree of cross-linking or the concentration of the gel. When they are suspended in a solvent, the gel particles absorb the solvent in amounts proportional to the degree of internal cross-linkage, forming pores of relatively uniform size (Williams, 1972). The solvent-saturated gel particles are then packed into a column. The area within the pore of the beads is termed the stationary phase, while the remaining are called the mobile phase (liquid phase). When a sample containing molecules of different sizes is applied onto the column, molecules that have dimensions larger than the largest pore size of the beads (i.e., above the exclusion limit) do not penetrate the gel particles and will pass through the mobile phase (i.e., liquid phase) outside the gel particles and emerge with the eluting solvent front. However, molecules smaller than the pore size of the gel particles can penetrate the beads to varying degrees depending on their sizes. Therefore, they will be eluted from the column in the order of their sizes. (For additional information see the Pharmacia booklet "Gel Filtration: Theory and Practice.")

Unlike ion-exchange chromatography (Section II-B, p. 9) in which the solvent (elutant) plays an active role in the separation, the elutant in gel filtrations passively carries the molecules down the column. Recovery, therefore, requires the use of a single buffer of fixed ionic strength, not gradient or step elutions. Equilibration in this instance is achieved by washing the column with 2–3 column volumes of the elution buffer immediately prior to loading the sample.

Because of their use in other sections of this manual, a few gel filtration systems warrant a brief description: Dextran, agarose and polyacrylamide, are used to form beads for gel filtration. Sephadex is a chromatography medium prepared by cross-linking dextran with epichlorohydrin. The large number of −OH groups carried by this carbohydrate makes it very hydrophobic, causing rapid swelling in water. Swelling is not effected by salts or detergents, however. Various degrees of cross-linking, and therefore fractionation ranges, are available. These are indicated by the G-type: type G-25 has a high degree of cross-linking and displays a fractionation range of $10^3 - 5 \times 10^3$ daltons; type G-50 is moderately cross-linked and has a fractionation range of $1.5 \times 10^3 - 3.0 \times 10^4$ daltons. Both G-25 and G-50 are useful for the separation of single nucleotides (such as radiolabeled precursors) from higher molecular weight oligonucleotides (see Sections XI, p. 59, and XII, p. 67, and Chapter 4, p. 157). Sephadex G-100 with a smaller degree of cross-linking separates molecules in the broad range of $4 \times 10^3 - 1.5 \times 10^5$ daltons. It can be used to separate oligonucleotides from polynucleotides (see Chapter 4, p. 157). Molecules with molecular weights greater than the upper limit of these ranges are excluded from the gel and elute with the void volume.

Agarose-based filtration systems are used in this manual as well. Unlike the dextran-based systems, where the gels are stabilized by covalent linkages, agarose-derived beads rely on hydrogen bonds for their stability. For this reason they are relatively temperature- (melts at 40°C) and chemical-sensitive. They do offer, however, an operating range in water of pH $4-13$ and swelling (i.e., porosity) is not significantly altered by high salt concentrations. Variation in pore size, and hence in fractionation ranges, is achieved by altering the concentration of agarose in the gel preparation. Bio-gel A-0.5M contains 10% agarose and displays a fractionation range of 10^4 to 5×10^5 daltons. It is used to separate oligonucleotide linker molecules from larger polynucleotides (see Chapter 4, p. 157). Sepharose-CL is agarose that has been cross-linked by treatment with 2,3-dibromopropanol under strongly alkaline conditions, thereby increasing its thermal (can be autoclaved) and chemical stability (Porath *et al.*, 1971). Sepharose CL-4B contains 4% agarose and has a fractionation range of 3×10^4 to 2.0×10^7. It can be substituted for Bio-gel A in the "cleaning-up" of reaction mixtures (see Chapter 4, p. 157).

The procedure outlined below refers specifically to Sephadex G-50 but the same steps may be followed in the preparation of any gel filtration column.

PROCEDURE

1. Prepare a suspension ("slurry") of Sephadex G-50 in TE buffer, pH 7.5. Allow the gel to hydrate ("swell") overnight at room temperature.

2. Use a disposable 10 ml plastic pipette to form the column (Figure 1.14). Cut off its cotton-plugged end and clamp the pipette to a support stand so that the newly opened end is uppermost.

3. Attach a $5-7$ cm length of 7 mm diameter plastic tubing to the lower end of the pipette (column exit). Attach a spring clamp to the tubing.

4. Insert a small wad of siliconized glass wool into the pipette (wearing disposable gloves for this step). With a second, smaller diameter pipette, push the glass wool to the bottom of the larger pipette. The glass wool will serve as a support for the Sephadex.

COMMENTS

1. The amount of dry gel needed to pack a column of known volume may be determined from the hydrated bed volume value supplied by the manufacturer. Because of spillage, repacking etc., it is wise to use two times as much buffer as needed to form the column.

2. If Sephadex is already presoaked, these columns may be easily prepared within 2 hours.

3. This will be used to control the flow of the column.

4. The glove prevents "staining" of the glass wool with skin oils, etc. Siliconization (part H, p. 100, this section) serves to prevent adhesion of DNA to the glass fibers. The glass wool should not be tightly packed, as this leads to slower flow rates.

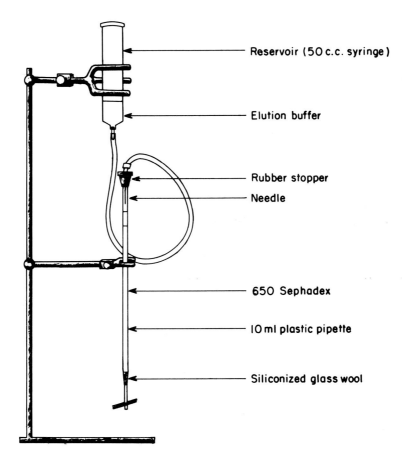

FIGURE 1.14 Gel filtration apparatus.

Labels in figure:
- Reservoir (50 c.c. syringe)
- Elution buffer
- Rubber stopper
- Needle
- 650 Sephadex
- 10 ml plastic pipette
- Siliconized glass wool

PROCEDURE

COMMENTS

5. Wet the inside of the 10 mL pipette and the glass wool with elution buffer (TE buffer, pH 7.5).

6. Drain the TE buffer through the column. Check the flow rate and adjust the glass-wool plug if necessary. Replace the spring clamp.

7. Fill the pipette with a slurry of medium grain Sephadex G-50 in elution buffer. Allow the Sephadex to settle.

7. The hydrated bed volume of Sephadex G-50 is 9–11 mL/g dry weight.

8. To further compact the Sephadex, open the column exit and allow buffer to run off.

9. Add more slurry. Repeat steps 6 and 7 until the column is approximately 18 cm in length.

9. The "0" mL mark on the pipette provides a convenient measure of the appropriate column height.

10. Use the barrel portion of a 50 mL syringe as a reservoir for the elution buffer. Attach it to a support stand at a position 2–3 cm above the top of the column.

10. If the elution buffer differs from the swelling buffer, the column should be equilibrated at this point by washing it with 2–3 column volumes of elution buffer.

11. Connect the reservoir to the upper end of the column (as indicated in Figure 1.14), using medium-gauge plastic tubing with disposable needles inserted at both ends. (Remove the metal portion from the reservoir needle). Close the tubing with a spring clamp.

12. Add buffer to the reservoir. After removing air from the system, connect the reservoir to the column by inserting the needle through the rubber stopper or cap used to seal the column.

12. When using a rubber cap to close the entrance to the column, make only one puncture. The caps are not self-sealing and a second hole will disrupt the self-contained nature of the system and make the flow rate difficult to control.

13. Prior to the addition of the sample (e.g., nick translation mixture) remove the column cap and drain the elution buffer until the meniscus reaches the Sephadex.

14. Carefully add the reaction solution and run it into the Sephadex. Add 500 μL of elution buffer and run the column until the meniscus reaches the Sephadex. Repeat once more.

14. Use plastic or siliconized glass pipettes for this.

15. Add 1 mL of elution buffer and connect the column with the buffer reservoir. Let the column flow.

16. Collect fractions of desired volume (0.5 mL) into 1.5 mL microfuge tubes.

17. Regenerate the column by washing it extensively with elution buffer.

17. If properly sealed after use, this column may be used repeatedly even after long intervals of storage.

NOTE: *If radioactive materials were separated, the washings should be disposed of as described in Section XII, p. 67.*

If the column is to be used for quantitative purposes such as molecular weight estimations, it must be calibrated using standardized molecular weight markers (available commercially). They should always contain one marker of molecular weight greater than the exclusion limit of the gel and several falling within its fractionation range. The void volume of a particular column may be determined with colored proteins (e.g., hemoglobin, ferritin) or blue dextran. Skewing or uneven migration of these colored bands indicates a lack of homogeneity within the column. Since this may affect molecular weight determinations, it should be corrected by repacking the column.

O. ACID FIXATION OF NUCLEIC ACIDS FOR LIQUID SCINTILLATION COUNTING

Liquid scintillation counting is a method of assessing the radioactivity of a preparation through the use of a solution of fluors and a photomultiplier tube. The scintillation solution converts the energy of the radioactive decay particle emitted by the isotope to light. The phototube responds to this light energy by producing a charge pulse that can be amplified and counted electronically. Typical scintillation solutions are toluene-based and contain PPO (2,5-diphenyloxazole) and POPOP (1,4-bis-2(5-phenyloxazolyl benzene) or bis-MSB (p-bis-(0-methylstyryl)-benzene) as primary and secondary fluors, respectively. Liquid scintillation counting is particularly useful for the detection of low-energy beta emitters (e.g., ^{3}H, ^{14}C). Certain high energy beta emitters (e.g., ^{32}P, ^{24}Na, ^{36}Cl) can be detected by liquid scintillation counting without the use of organic solvents and fluors. Rather, because their decay energy is sufficient to cause water to emit a bluish-white light (usually termed Cerenkov light), the immersion of these isotopes in this solvent is all that is required for detection. The radioisotope ^{32}P is detectable at about 40% efficiency by such **CERENKOV COUNTING** (for a general reference on radioisotope techniques, see Williams and Wilson, 1975).

In order to confirm the presence of a radioactive product in a particular solution, it must be separated from unincorporated isotopes and other extraneous materials. trichloroacetic acid (TCA)-mediated fixation of nucleic acids onto small pieces of filter paper provides one way of achieving this. After acid-treatment, non-fixed materials are removed from the paper by washing them in sequence with a series of general solvents. The radioactivity of the sample is then measured by liquid scintillation counting.

The filter paper counting method is simple, economical and lends itself to batch processing. In addition, spotting equal volumes of sample on each paper permits the estimation of the relative amount of radioactive material present in any particular fraction. This latter property, in combination with ultracentrifugation procedures, has been used to determine plasmid copy numbers (Zandvliet and Jansz, 1976).

PROCEDURE

1. Cut Whatman No. 1 filter paper into 1×1 or 2×2 cm squares. Number each square with a soft lead pencil, and arrange in sequence on a sheet of heavy-duty aluminum foil. (Make certain that the papers do not touch each other.)

2. Spot $5-50$ µL aliquots from each fraction onto the filter papers using a plastic-tipped micropipetter. Air dry or place in a $50-60°C$ oven until dry.

3. Place the dried papers in a beaker and cover with 200 ml of *cold* 10% TCA. Soak the papers at 4°C for 30 minutes. Change the TCA every 10 minutes. Collect discarded TCA in a screw-capped glass bottle, then dispose of it as outlined for radioactive waste in Section XII, p. 67.

4. Wash the TCA-treated filter papers with two changes of 100 mL ethanol followed by two changes of acetone.

5. Air dry the filter paper discs. Place the discs in separate counting vials and cover them with scintillation fluid.

6. Cap the vials and count the radioactivity in a liquid scintillation counter.

COMMENTS

1. Whatman No. 3MM cellulose discs (23 mm) may also be used. The size of the filter papers is determined by the volume of sample to be spotted onto it. With the larger squares more extensive folding is needed to fit them into the counting vials. Because less area is exposed to scintillation solution, counting will be less efficient and greater volumes of scintillation solution must be used.

2. The volume of sample used depends on the kind and amount of isotope used in the initial labeling.

3. Most macromolecules are fixed to cellulose by cold TCA. When nucleoside triphosphates are used for labeling, considerable binding of these compounds to the filter paper may occur. To inhibit this, supplement the TCA with 0.01 mM EDTA and 10% sodium pyrophosphate.

4. Ethanol removes water and all nonfixed alcohol-soluble materials. Acetone (or ether) removes ethanol and alcohol-insoluble materials.

Treat the ethanol/acetone as radioactive waste (see Section XII, p. 67).

6. After removal of the filter paper, the scintillation fluid in nonpeak vials may be reused. Background levels should be ascertained after two uses. Plastic counting vials are of limited utility in this regard because of their permeability to toluene.

λ STOCK CULTURE

Bacteriophage are propagated by repeated cycles of infection and lysis of their host cell. Phage lambda (λ) adsorbs to specific receptor sites on the bacterial cell wall; infection begins when its DNA enters the cell. Viral genes are subsequently expressed and direct the synthesis of molecules required for phage reproduction, assembly of new phage particles, and finally, for the disruption of the bacterial cell wall (lysis). Cells infected by a single phage particle may release several hundred copies of the phage upon cell lysis. See Lewin (1977) and Campbell (1983) for more detailed information on the life cycle, genetics and propagation of λ phage. A method for preparing stock cultures of phage λ is outlined below.

PROCEDURE

1. Inoculate the appropriate host cells in 1 mL of B or LB broth and incubate overnight at 37°C.

2. Dilute the overnight culture 1:20 in to fresh broth and incubate at 37°C for 2−3 hours.

3. Add 0.2 mL of sterile 1.0M $MgSO_4$ to the bacterial culture. Then mix 0.1 mL of bacteriophage λ with 0.2 mL of bacterial culture. Incubate for 10−15 minutes at 37°C.

4. Add 2.5 mL of molten top agar (45°C) to the preadsorption mixture, gently mix and immediately pour over the surface of a bottom agar plate.

5. Allow the top agar to solidify, then incubate the plate face up at 37°C for 8−18 hours.

6. Use a "hockey stick" (L-shaped glass rod) to scrape the top agar into a sterile centrifuge tube.

7. Add 2 mL of sterile broth (B or LB) to each centrifuge tube and mix vigorously.

COMMENTS

1. Appropriate host cells for λ include *E. coli* strains DP50, K802, SF8 and HB101.

B Broth:

> 10 g Bacto tryptone
> 8 g NaCl
> 1 L distilled H_2O

LB Broth:

> 10 g Bacto tryptone
> 5 g Bacto yeast extract
> 10 g NaCl
> 1 L distilled H_2O

2. The host cell count should be $3-5 \times 10^8$ colony forming units (cfu) per mL.

3. A final concentration of 0.01M $MgSO_4$ is necessary for the preadsorption of bacteriophage λ. The concentration of phage should be approximately $1-2 \times 10^6$ plaque-forming units (pfu)/mL, (*λ:E. coli* ≈ *1:100*).

4. *TOP AGAR:* Same as B or LB broth, except 0.7−0.8% of agar is added (i.e., 7−8 g of agar per L of broth).

BOTTOM AGAR: same as top agar, except 1.5% of agar is added (i.e., 15 g of agar per L of broth).

The bottom agar should be freshly prepared. Be careful not to introduce any air bubbles.

5. After this period of incubation most cells should be lysed (i.e., no discrete plaques will be seen).

7. It is possible to pool the scrapings from two plates. In this instance add 4 mL of broth.

At this step, 5–10 drops of chloroform can be added to lyse the bacterial cells. However, the centrifuge tube must be chloroform resistant (e.g., glass or cellulose nitrate).

8. Pellet the cell debris by centrifugation at 2000 rpm at room temperature for 10–15 minutes. Remove the supernatant and add 2 drops of chloroform. Store at 4°C.

8. Chloroform is added to keep the lysate sterile and to kill any remaining viable bacteria.

9. Titrate the phage preparation using serial dilutions of the lysate made in either B or LB broth.

9. *TITRATION OF PHAGE:* Dilute the lysate in broth to 10^{-4}, 10^{-5}, 10^{-6}, 10^{-7} and 10^{-8}. Add 0.1 mL of each dilution to 0.2 mL of the host bacterium (in exponential phase). Then proceed as in steps 4 and 5.

10. Count the number of plaques per dilution and calculate the titer in plaque-forming units (pfu) per mL.

Q. PREPARATION AND ASSAY OF THE FILAMENTOUS PHAGE M13

Bacteriophage M13 and its derivatives (M13mp7, M13mp8 and M13mp9) are filamentous, male-specific (infect only F^+, F' or Hfr strains), single-stranded, DNA bacteriophages similar to other filamentous phages (e.g., fd and f1) (Messing and Vieira, 1982; Zinder and Boeke, 1982). M13 does not lyse its host, rather it is extruded from infected cells without damaging them as they continue to grow and divide (Denhardt *et al.*, 1978). Upon infection, the single-stranded phage DNA is converted into a double-stranded replicative form (RF) which then goes through several rounds of replication to increase its copy numbers (100–200 per cell). These amplified replicative forms serve as intermediates in the production of progeny (+ strand) single-stranded DNA. The latter are packaged into a protein coat and then are continuously released from the cell. M13 infected cells grow very slowly, and, in contrast to the clear plaques obtained with λ or T_4 bacteriophages, produce turbid plaques.

1. Inoculate 1 mL of 2xYT medium with an appropriate host (e.g., *E. coli* K38, JM101) and incubate at 37°C overnight.

1. 2xYT Medium:

16 g Bacto tryptone
10 g Bacto yeast extract
 5 g NaCl
1 L distilled H_2O

2. Dilute the overnight culture 20–50-fold with fresh 2xYT medium and incubate at 37°C for 1–2 hours.

3. Using sterile techniques, infect 2 mL of host cell culture with a single plaque.

3. Using a sterile Pasteur pipette, remove a single plaque from the surface of the top agar and transfer it to the host cell culture.

4. Incubate at 37°C for 5−8 hours with aeration (100−200 rpm).

5. Pellet the host bacteria by centrifugation at 7000 rpm for 10 minutes at room temperature.

6. Heat the supernatant to 65°C for 30 minutes to kill remaining host cells. The phage titer should be $10^{11}-10^{12}$ pfu/mL.

6. The phage can be concentrated by precipitation with polyethylene glycol, molecular weight, 6000, (PEG6000) in the presence of 0.5M NaCl (Yamamoto *et al.*, 1970).

Concentration of Phage with PEG6000:

a. Prepare a solution containing 40% PEG6000 and 2.5M NaCl.

b. Add 0.6 mL of the above solution to 2 mL of the phage lysate, such that the final concentration of PEG is about 10% and the concentration of NaCl is 0.5M. If the volume of the phage lysate is above 50 mL, solid PEG6000 and NaCl may be added directly. In this case, the PEG and NaCl are dissolved by stirring with a magnetic stirrer at room temperature.

c. Let stand at 4°C for 1−2 hours.

d. Harvest the precipitated phage particles by centrifugation at 10,000 g for 10 minutes at 4°C.

e. Discard the supernatant, invert the centrifuge tube and air dry.

f. Resuspend the precipitated phage in appropriate volume of broth or saline.

g. The resuspended phage can be further purified by CsCl gradient centrifugation (Section II, p. 7).

7. The titer of the phage can be assayed using the same procedure as described in propagation of bacteriophage λ (see part P, p. 108, this section).

R. TESTS FOR DETECTING β-LACTAMASE-PRODUCING CELLS

Many cloning vectors carry genes that determine resistance to penicillin-group antibiotics. Resistance is mediated by the production of an enzyme (β-lactamase) which hydrolyzes the β-lactam bond of the penicillin molecule. Several types of β-lactamase are produced by bacteria and their plasmids (Matthew *et al.*, 1979; Sykes and Matthew, 1976); however, the structural gene for the TEM-type β-lactamase *(bla)*, which is specified by many plasmids (Heffron *et al.*, 1975), is a gene that has been incorporated into several cloning vectors (e.g., pBR322; Bolivar *et al.*, 1977a). The *bla* gene is expressed in many hosts, including *S. cerevisiae* (Hollenberg, 1982). Several tests have been developed to detect β-lactamases from a variety of sources. In the following section two procedures and their modifications are described, the chromogenic cephalosporin test and iodometric assay.

METHOD 1: CHROMOGENIC CEPHALOSPORIN TEST

The chromogenic cephalosporin, nitrocefin, undergoes a color change when hydrolyzed by β-lactamases from a variety of sources (O'Callaghan *et al.*, 1972). An advantage to this test is that colonies treated with subinhibiting concentrations of the compound remain viable and can be subcultured directly. In some areas, nitrocefin powder cannot be obtained; thus a paper disc variation to the method must be used.

PROCEDURE	COMMENTS

a. AGAR PLATE TEST:

1. Spread or patch (i.e., pick individual colonies) the cells to be tested on a clear enriched agar medium (pH 7.0–7.5). Incubate until reasonable growth is obtained.

1. Use of a dark-colored medium such as McConkey agar may make the chromogenic cephalosporin color changes difficult to observe.

2. Prepare a working solution of the antibiotic nitrocefin.

2. *CAUTION:* *Avoid inhalation and contact with eyes and skin of nitrocefin powder.*

Working Solution: *Nitrocefin*

Dissolve 5 mg of the powder in 0.5 mL of dimethyl sulphoxide. Add 9.5 mL of phosphate buffer (0.1M, pH 7.0) and mix.

Store in a brown screw-cap bottle at 4°C. The solution is stable for long periods when refrigerated but rapidly becomes inactivated at room temperature.

0.1M Phosphate Buffer:

a. Prepare stock solutions of 0.2M dibasic sodium phosphate (A) and 0.2M monobasic sodium phosphate (B).

b. Mix solutions A and B as follows:

pH at 25°C	Solution A ml	Solution B ml
6.0	43.85	6.15
7.0	19.50	30.50

(The pH 6.0 buffer is used in Method 2 of this section).

c. Dilute the mixtures to 100 mL. The buffer will be stable for several weeks.

3. Using a Pasteur pipette, place one or two drops of the nitrocefin solution on the bacterial growth.

3. Alternatively, the bacterial growth may be suspended in 2 drops of nitrocefin solution placed either on a slide or in a microtitre well. All tests should be conducted using positive and negative control strains.

4. A β-lactamase producing colony produces a red color almost immediately; with a β-lactamase negative colony, the nitrocefin solution remains yellow.

4. With most β-lactamase-producing isolates the solution turns red immediately. Some cultures (e.g., yeast) may be weak β-lactamase producers, and a longer period of incubation (at room temperature) may be required.

b. PAPER DISC METHOD

In some areas, it is difficult to obtain nitrocefin powder. However, discs impregnated with this compound, called Cefinase discs (available from Becton Dickinson and Company), have recently become available. The procedure recommended by the manufacturer is as follows:

1. Using sterile forceps, remove the number of discs needed and place them in an empty Petri plate.

1. Other companies have developed commercial tests for screening for β-lactamase production (e.g., Bethesda Research Laboratories, *Ampscreen*™).

2. Moisten each disc with 1−2 drops of sterile water.

3. Smear several pure colonies onto the surface of the disc with a sterile loop or toothpick.

4. If an isolate produces β-lactamase, the disc will turn red at the inoculation site within five minutes. Otherwise, it remains yellow.

METHOD 2:   IODOMETRIC TEST

The iodometric test is an inexpensive, simple, and reproducible test for detecting β-lactamase-producing (or nonproducing) colonies. It was originally developed by Perret (1954) and has since been modified by many workers for a variety of purposes (e.g., Novick, 1962; Ross and O'Callaghan, 1975; Jorgensen *et al.*, 1977). The product of β-lactamase hydrolysis (e.g., penicilloic acid in the case of penicillin) is oxidized by iodine. Thus, in the presence of starch-iodine complexes, which are blue-black, the oxidation process removes iodine from the starch-iodine complex, thereby decolorizing it.

a. AGAR PLATE METHOD

1. Incorporate starch into an agar plate such that the final concentration is 0.2%.

1. Some media (e.g., GC medium base, Difco) contain enough starch and do not require an additional supplement.

For *Sacchromyces cerevisiae*, the growth medium may contain:
  0.65% yeast nitrogen base
  0.1%   glucose
  0.2%   starch
  0.02M phosphate buffer (pH 6.5)
    2%   agar

2. Spread or patch the test cells on the starch plate. Incubate until reasonable growth is obtained.

2. **PATCHING** refers to the subculture of colonies, as pin prick inoculi, on a plate. As many as 100 colonies may be inoculated onto a single plate.

PROCEDURE	COMMENTS

3. Prepare the **PENICILLIN STOCK SOLUTION** by adding the appropriate amount of pencillin powder to 0.1M phosphate buffer, pH 6.0, to give a final concentration of 6000 µg/mL.

3. 1 unit of penicillin = 0.6 µg

Solutions should either be used immediately or stored in 10 mL aliquots at −20°C. Thawed aliquots should not be refrozen.

4. Prepare the Iodine Stock Solution.

4. Iodine Stock Solution:

   2.03 g iodine
  52.02 g potassium iodide
100.00 mL distilled water

Dissolve and store in a brown bottle.

The solution can be kept indefinitely.

5. Prepare the Iodine-Penicillin Test Solution.

5. Iodine-Penicillin Test Solution:

Combine 1.7 mL iodine stock solution with 10 mL of the penicillin stock solution. Mix.

The solution should be freshly prepared.

6. Flood the plate from step 2 with 3−5 mL of the iodine-penicillin test solution and aspirate excess solution with a Pasteur pipette. The medium will immediately turn blue-black or deep brown. The presence of β-lactamase-producers is indicated by the decoloration of the colonies followed by the formation of a colorless surrounding zone. With time this zone will gradually expand in size. Colonies which do not produce β-lactamase do not yield colorless zones and appear yellow in color.

6. The test solution might be included in a soft agar overlay.

b. FILTER PAPER METHOD

1. Impregnate filter paper (Whatman No. 1) with freshly prepared penicillin stock solution and a 0.2% starch solution. Air dry.

1. 0.2% STARCH SOLUTION: Add 2.0 grams of soluble starch to 100 mL distilled water. Dissolve by boiling. This solution must be freshly prepared. Filters may be stored in a sealed desiccator at 4°C for up to two weeks or at −20°C for longer periods.

2. Add 1 mL of the iodine stock solution to 9 mL of distilled water.

3. Moisten the impregnated filters with the diluted iodine solution. The paper will turn dark brown in color.

4. Place the filter paper directly on the surface of a previously inoculated enriched agar plate.

4. Alternatively, colonies may be picked from the agar plate with a sterile toothpick and streaked on the paper.

5. β-lactamase-producing colonies are readily detected by the gradual development of a surrounding zone of decolorization (i.e., paper turns whitish).

XVIII. References

Aaij, C., and P. Borst. 1972. The gel electrophoresis of DNA. *Biochem. Biophys. Acta.* 269:192−200.

Abram, M.B., H.T. Ballantine, G.R. Dunlop, B.K. Jacobson, J.J. Moran, A.G. Motulsky, D.B. Rahi, S. Siegel, L. Smith, K. Toma, and C.J. Walker. 1982. "Splicing life. The social and ethical issues of genetic engineering with human beings." President's Commission for the Study of Ethical Problems in Medicine and Biomedical and Behavioral Research. U.S. Government Printing Office, Washington, D.C. (LC#83-600500).

Alwine, J.C., D.J. Kemp, and G.S. Stark. 1977. Method for detection of specific RNAs in agarose gels by transfer to diazobenzyloxymethyl paper and hybridization with DNA probes. *Proc. Natl. Acad. Sci. U.S.A.* 74:5350−5354.

Alwine, J.C., D.J. Kemp, B. Parker, J. Reiser, J. Renart, G. Start, and G.M. Wahl. 1979. Detection of specific RNAs or specific fragments of DNA by fractionation in gels and transfer to diazobenzyloxymethyl paper, 220−228. In R. Wu (ed.), Methods in Enzymology, vol. 68. Academic, New York.

Andrews, A.T. 1981. Electrophoresis: theory, techniques and biochemical and clinical applications. Clarendon, Oxford.

Appleyard, R.K. 1954. Segregation of new lysogenic types during growth of a doubly lysogenic strain derived from *E. coli* K12. *Genetics* 39:440−444.

Arber, W. 1974. DNA modification and restriction, 1−37. In W.E. Cohen (ed.), Progress in Nucleic Acid Research and Molecular Biology. Academic, New York.

Aviv, H., and P. Leder. 1972. Purification of biologically active globin messenger RNA by chromatography on oligothymidylic acid-cellulose. *Proc. Natl. Acad. Sci. U.S.A.* 69:1408−1412.

Avni, H., and A. Markovitz. 1979. Characterization of a mini-ColEI cloning vector. *Plasmid* 2:225−236.

Bagdasarian, M., R. Lurz, B. Ruchert, F.C.H. Franklin, M.M. Bagdasarian, J. Frey, and K.N. Timmis. 1981. Specific purpose plasmid cloning vectors II. Broad host range, high copy number, RSF1010-derived vectors, and a host-vector system for gene cloning in *Pseudomonas*. *Gene* 16:237−247.

Bahl, C.P., K.J. Marians, R. Wu, J. Stawinsky, and S.A. Narang. 1976. A general method for inserting specific DNA sequences into cloning vehicles. *Gene* 1:81−92.

Bahl, C.P., and R. Wu. 1978. Cloned seventeen nucleotide-long synthetic lactose operator is biologically active. *Gene* 3:123−134.

Baltimore, D. 1970. RNA-dependent DNA polymerase in virions of RNA tumor viruses. *Nature* 226:1209−1211.

Banner, D.B. 1982. Recovery of DNA fragments from gels by transfer to DEAE-paper in a electrophoresis chamber. *Anal. Biochem.* 125:139−142.

Barnes, W.M. 1980. DNA cloning with single-stranded phage vectors, 185−200. In J. Setlow and A. Hollaender (ed.), Genetic Engineering: Principles and Methods, vol. II. Plenum, New York.

Benton, W.D., and R.W. Davis. 1977. Screening λgt recombinant clones by hybridization to single plaques *in situ*. *Science* 196:180−182.

Bernard, H.U., and D.R. Helinski. 1980. Bacterial plasmid cloning vehicles, p. 133−167. In J. Setlow and A. Hollaender, (ed.), Genetic Engineering: Principles and Methods, vol. II. Plenum, New York.

Best, A.N., D.P. Allison, and G.P. Novelli. 1981. Purification of supercoiled DNA of plasmid ColE1 by RCP-5 chromatography. *Anal. Biochem.* 114:235−243.

Betlach, M., V. Hershfield, L. Chow, W. Brown, H.M. Goodman, and H.W. Boyer. 1976. A restriction endonuclease analysis of the bacterial plasmid controlling the *Eco*RI restriction and modification of DNA. *Fed. Proc., Fed. Am. Soc. Exp. Biol.* 35:2037−2043.

Bezanson, G.S., R. Khakhria, and R. Lacroix. 1982. Involvement of plasmids in determining bacteriophage sensitivity in *Salmonella typhimurium:* genetic and physical analysis of phagovar 204. *Can. J. Microbiol.* 28:993−1001.

Birnboim, H.C., and J. Doly. 1979. A rapid alkaline extraction procedure for screening recombinant plasmid DNA. *Nucl. Acids. Res.* 7:1513–1523.

Birnboim, H.C. 1983. A rapid alkaline extraction method for the isolation of plasmid DNA, 243–255. In R. Wu, L. Grossman, and K. Moldave (ed.), Methods in Enzymology, vol. 100. Academic, New York.

Blattner, F.R., W. Williams, A. Blechl, K. Denneston-Thompson, H. Faber, L.A. Furlong, D. Greenwald, D. Kiefer, D. Moore, J. Schumm, E. Sheldon, and O. Smithes. 1977. Charon phages: safer derivatives of bacteriophage lambda for DNA cloning. *Science* 196:167–169.

Blattner, F.R., A.E. Blechl, K. Denneston-Thompson, H.E. Faber, J.E. Richards, J.L. Slightom, R.W. Tucker, and O. Smithies. 1978. Cloning human fetal γ globin and mouse α type globin DNA preparation and screening of shotgun collections. *Science* 202:1279–1284.

Blin, N., A.V. Gabain, and H. Bujard. 1975. Isolation of large molecular weight DNA from agarose gels for further digestion by restriction enzymes. *FEBS Lett.* 53:84–86.

Bolivar, F., and K. Backman. 1979. Plasmids of *Escherichia coli* as cloning vectors, 245–267. In R. Wu (ed.), Methods in Enzymology, vol. 68. Academic, New York.

Bolivar, F., R. Rodriquez, M. Betlach, and H. Boyer. 1977a. Construction and characterization of new cloning vehicles I: ampicillin resistant derivatives of plasmid pMB9. *Gene* 2: 75–93.

Bolivar, F., R. Rodriguez, P. Greene, M. Betlach, H. Heyneker, H. Boyer, J. Croza, and S. Falkow. 1977b. Construction and characterization of new cloning vehicles. II. A multipurpose cloning system. *Gene* 2:95–113.

Bollum, F.J.. 1962. Oligonucleotide-primed reactions catalysed by calf thymus polymerase. *J. Biol. Chem.* 237:1945–1962.

Bonner, W.M. and R.A. Laskey. 1974. A film detection method for tritium-labeled proteins and nucleic acids in polyacrylamide gels. *Eur. J. Biochem.* 46:83–88.

Boyer, H.W. 1971. DNA restriction and modification mechanisms in bacteria. *Ann. Rev. Microbiol.* 25:153–176.

Broach, J.R., J.N. Strathen, and J.B. Hicks. 1979. Transformation in yeast: development of a hybrid cloning vector and isolation of the CAN1 gene. *Gene* 8:121–133.

Broda, P. 1979. Plasmids. Freeman, San Francisco.

Brunk, C.F., and L. Simpson. 1977. Comparison of various ultraviolet sources for fluorescent detection of ethidium bromide-DNA complexes in polyacrylamide gels. *Anal. Biochem.* 82:455–462.

Brutlag, D., K. Fry, T. Nelson, and P. Hung. 1977. Synthesis of hybrid bacterial plasmids containing highly repeated satellite DNA. *Cell* 10:509–519.

Bukhari, A.I., J.A. Shapiro, and S.L. Adhya, eds. 1977. DNA insertion elements, plasmids and episomes. Cold Spring Harbor Laboratory, Cold Spring Harbor, New York.

Bunemann, H., and W. Müller. 1978. Affinity Chromatography, 353–356. In O. Hoffman-Ostenhaf (ed.), Affinity Chromatography: Proceedings at an International Symposium Held in Vienna. Pergamon, New York.

Burckhardt, J., and M.L. Birnstiel. 1978. Analysis of histone messenger RNA of *Drosophila melanogaster* by two-dimensional gel electrophoresis. *J. Mol. Biol.* 118:61–79.

Burnette, W.N. 1981. Western blotting: electrophoretic transfer of proteins from sodium dodecyl sulfate polyacrylamide gels to unmodified nitrocellulose and radiographic detection with antibody and radioiodinated protein A. *Anal. Biochem.* 112:195–203.

Calos, M.P., J.S. Lebknowski, and M.R. Botcham. 1983. High mutation frequency of DNA transfected into mammalian cells. *Proc. Natl. Acad. Sci. U.S.A.* 80:3015–3019.

Campbell, A. 1983. Bacteriophage λ, 65–103. In J.A. Shapiro (ed.), Mobile Genetic Elements. Academic, New York.

Casse F., C. Boucher, J.S. Julliot, M. Michel, and J. Dénairié. 1979. Identification and characterization of large plasmids in *Rhizobium meliloti* using agarose gel electrophoresis. *J. Gen. Microbiol.* 113:229–242.

Chaconas, G., and J.H. Van de Sande. 1980. 5' ^{32}P labeling of RNA and DNA restriction fragments, 75—85. In L. Grossman, and K. Moldave (ed.), Methods in Enzymology, vol. 65. Nucleic Acids, Part One. Academic, New York.

Chakrabarty, A.M., J.R. Mylroie, D.A. Friella, and J.G. Vacca. 1975. Transformation of *Pseudomonas putida* and *Escherichia coli* with plasmid-lined drug-resistance factor DNA. *Proc. Natl. Acad. Sci. U.S.A.* 72:3647—3651.

Challberg, M.D., and P.T. Englund. 1980. Specific labeling of 3' termini with T4 DNA polymerase, 34—43. In L. Grossman, and K. Moldave (ed.), Methods in Enzymology, vol. 65. Academic, New York.

Chang, A.C.Y., and S.N. Cohen. 1978. Construction and characterization of amplifiable multicopy DNA cloning vehicles derived from the P15A cryptic miniplasmid. *J. Bacteriol.* 134:1141—1156.

Chang, A.C.Y., J.H. Nunberg, R.J. Kaufman, H.A. Erlich, R.T. Schimke, and S.N. Cohen. 1978. Phenotypic expression in *E. coli* of a DNA sequence coding for mouse dihydrofolate reductase. *Nature* 275:617—624.

Chang, L.M., and F.J. Bollum. 1971. Deoxynucleotide-polymerizing enzymes of calf thymus gland. V. Homogeneous terminal deoxynucleotidyl transferase. *J. Biol. Chem.* 246:909—916.

Chang, S., O. Gray, D. Ho., J. Koyer, S.-V. Chang, J. McLaughlin, and D. Mark. 1982. Expression of eucaryotic genes in *B. subtilis* using signals of *pen*P, 159—169. In A.T. Ganeson, S. Chang, and J. Hoch (ed.), Molecular Cloning and Gene Regulation in Bacilli. Academic, New York.

Chou, S., and T.C. Merigan. 1983. Rapid detection and quantitation of human cytomegalovirus in urine through DNA hybridization. *New Engl. J. Med.* 308:921—925.

Clayton, D.A., R.W. Davis, and J. Vinograd. 1970. Homology and structural relationships between the dimeric and monomeric circular forms of nitrochondrial DNA from human leukemic leukocytes. *J. Molec. Biol.* 47:137—153.

Clewell, D.B., and D.R. Helinski. 1969. Supercoiled circular DNA-protein complex in *Escherichia coli*: purification and induced conversion to an open circular DNA form. *Proc. Natl. Acad. Sci. U.S.A.* 62:1159—1166.

Clewell, D.B., and D.R. Helinski. 1972. Effect of growth conditions on the formation of the relaxation complex of supercoiled ColEI deoxyribonucleic acid and protein in *Escherichia coli*. *J. Bacteriol.* 110:1135—1146.

Cohen, S.N., A.C.Y. Chang, and L. Hsu. 1972. Nonchromosomal antibiotic resistance in bacteria: genetic transformation of *Escherichia coli* by R-factor DNA. *Proc. Natl. Acad. Sci. U.S.A.* 69:2110—2114.

Collins, J., and B. Hohn. 1978. Cosmids: a type of plasmid gene-cloning vector that is packagable *in vitro* in bacteriophage λ heads. *Proc. Natl. Acad. Sci. U.S.A.* 75:4242—46.

Colman, A., M.J. Byers, S.B. Primrose, and A. Lyons. 1978. Rapid purification of plasmid DNAs by hydroxyapatite chromatography. *Eur. J. Biochem.* 91:303—310.

Currier, T.C., and E.W. Nester. 1976. Isolation of covalently closed circular DNA of high molecular weight from bacteria. *Anal. Biochem.* 76:431—441.

Curtiss III, R., M. Inoue, D. Pereira, J.C. Hsu, L. Alexander, and L. Rock. 1977. Construction and use of safer bacterial host strains for recombinant DNA research, 99—114. In W.A. Scott, and R. Werner (ed.), Molecular Cloning of Recombinant DNA. Academic, New York.

Dagert, M., and S.D. Ehrlich. 1979. Prolonged incubation in calcium chloride improves the competence of *Escherichia coli* cells. *Gene* 6:23—28.

Daigneault, R., G. Bellmare, and G.H. Cousineau. 1971. Effect of various inhibitors on the activity of sea urchin ribonucleases. *Lab. Pract.* 20:487—488.

Davis, B., and E. Mingioli. 1950. Mutants of *Escherichia coli* requiring methionine or vitamin B12. *J. Bacteriol.* 60:17—28.

Davies, R.W. 1982. DNA sequencing, 117—172. In D. Rickwood and B.D. Hames (ed.), Gel Electrophoresis of Nucleic Acids: a practical approach. IRL, Oxford.

Deininger, P.L. 1983. Approaches to rapid DNA sequence analysis. *Anal. Biochem.* 135: 247−263.

Denhardt, D.T. 1966. A membrane-filter technique for the detection of complementary DNA. *Biochem. Biophys. Res. Commun.* 23:641−646.

Denhardt, D.T., D. Dressler, and D.S. Ray (ed.), 1978. The single-stranded DNA phages. Cold Spring Harbor Laboratory, Cold Spring Harbor, New York.

DeWachter, R., and W. Fiers. 1972. Preparative two-dimensional polyacrylamide gel electrophoresis of ^{32}P-labeled RNA. *Anal. Biochem.* 49:184−197.

DeWachter, R., and W. Fiers. 1982. Two dimensional gel electrophoresis of nucleic acids, 77−116. In D. Rickwood, and B.D. Holmes (ed.), Gel Electrophoresis of Nucleic Acids. IRL, Oxford.

Dillon, J.R., M. Pauzé, and K.-H. Yeung. 1983. Spread of penicillinase-producing and transfer plasmids from the gonococcus to *Neisseria meningitidis*. *Lancet* 1:779−781.

DiMais, D., R. Treesman, and T. Maniatis. 1982. Bovine papillomavirus vector that propagates as a plasmid in both mouse and bacterial cells. *Proc. Natl. Acad. Sci. U.S.A.* 79:4030−4034.

Drouin, J. 1980. Cloning of human mitochondrial DNA in *Escherichia coli*. *J. Mol. Biol.* 140:15−34.

Dubneau, D., T. Gryczan, S. Contente, and A.G. Shwakumar. 1980. Molecular Cloning in *Bacillus subtilis*, 115−131. In J.K. Setlow, and A. Hollaender (ed)., Genetic Engineering: Principles and Methods, vol. 2. Plenum, New York.

Dugaiczyk, A., H.W. Boyer, and H.M. Goodman. 1975. Ligation of *Eco*RI endonuclease-generated DNA fragments into linear and circular structures. *J. Mol. Biol.* 96:171−184.

Edman, J.C., R. Hallewell, R. Valenzuela, H. Goodman, and W. Rutter. 1981. Synthesis of hepatitis B surface and core antigens in *E. coli*. *Nature* 291:503−506.

Efstratiadis, A., and L. Villa-Komaroff. 1979. Cloning of double-stranded cDNA, 15−36. In J.K. Setlow and A. Hollaender (ed.), Genetic Engineering: Principles and Methods, vol. 1. Plenum, New York.

Ehrlich, M., and R.-H. Wang. 1981. 5-methylcytosine in eukaryotic DNA. *Science* 212:1350−1357.

Ehrlich, S.D., B. Niaudet, and B. Michel. 1982. Use of plasmids from *Staphylococcus aureus* for cloning of DNA in *Bacillus subtilis*, 19−29. In P.H. Hofschneider, and W. Goebel (ed.), Gene cloning in organisms other than *E. coli*. Springer-Verlag, Berlin.

Finkelstein, M., and R. Rownd. 1978. A rapid method for extracting DNA from agarose gels. *Plasmid* 1:557−562.

Fischer, S.G., and L.S. Lerman. 1979a. Two-dimensional electrophoretic separation of restriction enzyme fragments of DNA, 183−191. In R. Wu (ed.), Methods in Enzymology, vol. 68. Recombinant DNA. Academic, New York.

Fischer, S.G., and L.S. Lerman. 1979b. Length-dependent separation of DNA restriction fragments in two-dimensional gel electrophoresis. *Cell* 16:191−200.

Flores, J., R. Purcell, I. Perez, R. Wyatt, E. Boeggeman, M. Sereno, L. White, R. Chanock, and A. Kapikian. 1983. A dot hybridization assay for detection of rotavirus. *Lancet* 1:555−559.

Frei, E., A. Levy, P. Gowland, and M. Noll. 1983. Efficient transfer of small DNA fragments from polyacrylamide gels to diazo or nitrocellulose paper and hybridization, 309−326. In R. Wu, L. Grossman, and K. Moldave (ed.), Methods in Enzymology, vol. 100. Recombinant DNA. Academic, New York.

Freter, R. 1978. Possible effects of foreign DNA on pathogenic potential and intestinal proliferation of *Escherichia coli*. *J. Infect. Dis.* 137:624−629.

Fuchs, R., and R. Blakesley. 1983. Guide to the use of type II restriction endonucleases, 1−38. In R. Wu, L. Grossman, and K. Moldave (ed.), Methods in Enzymology, vol. 100. Recombinant DNA. Academic, New York.

Gaal, Ö, G.A. Medgyesi, and L. Vereczkey. 1980. Electrophoresis in the separation of biological molecules. Wiley, New York.

Gagnon, C., and G. de Lamirande. 1973. A rapid and simple method for the purification of rat liver RNase inhibitor. *Biochem. Biophys. Res. Commun.* 51:580–586.

Ganesan, A.T., S. Chang, and J.A. Hoch. 1982. Molecular Cloning and Gene Regulation in Bacilli. Academic, New York.

Garrels, J.I. 1983. Quantitative two-dimensional gel electrophoresis of proteins, 411–423. In R. Wu, L. Grossman, and K. Moldave (ed.), Methods in Enzymology, vol. 100. Recombinant DNA, Part B. Academic, New York.

Georges, M.C., I.K. Waschsmuth, K.A. Birkness, S.L. Moseley, and A.J. Georges. 1983. Genetic probes for enterotoxigenic *Escherichia coli* isolated from childhood diarrhea in the central African Republic. *J. Clin. Microbiol.* 18:199–202.

Gergen, J.P., R.H. Stern, and P.C. Wensink. 1979. Filter replicas and permanent collections of recombinant DNA plasmids. *Nucl. Acids Res.* 7:2115–2136.

Gilroy, T.E., and C.A. Thomas Jr. 1983. The analysis of some new *Drosophila* repetitive DNA sequences isolated and cloned from two-dimensional agarose gels. *Gene* 23:41–51.

Goddard, J.M., D. Caput, S.R. Williams, and D.W. Martin Jr. 1983. Cloning of human purine-nucleoside phosphorylase cDNA sequences by complementation in *Escherichia coli.* *Proc. Natl. Acad. Sci. U.S.A.* 80:4281–4285.

Goebel, W. 1970. Studies on extrachromosomal elements. Replication of the colicinogenic factor ColE1 in two temperature sensitive mutants of *Escherichia coli* defective in DNA replication. *Eur. J. Biochem.* 15:311–320.

Goeddel, D., D. Kleid, F. Bolivar, H. Heynekler, D. Yansara, R. Crea, T. Hirose, A. Drasjewski, K. Itakura, and A. Reggs. 1979. Expression in *Escherichia coli* of chemically synthesized genes for human isolation. *Proc. Natl. Acad. Sci. U.S.A.* 76:106–110.

Gorbach, S.L. 1978. Recombinant DNA: an infectious disease perspective. *J. Infect. Dis.* 137:615–623.

Grierson, D. 1982. Gel electrophoresis of RNA, 1–38. In D. Rickwood, and B.D. Hames, (ed.), Gel Electrophoresis of Nucleic Acids: a practical approach. IRL, Oxford.

Grunstein, M. and D.S. Hogness. 1975. Colony hybridization: a method for the isolation of cloned DNAs that contain a specific gene. *Proc. Natl. Acad. Sci. U.S.A.* 72:3961–3965.

Gryczan, T.J., S. Contente, and D. Dubnau. 1978. Characterization of *Staphylococcus aureus* plasmids introduced by transformation into *Bacillus subtilis. J. Bacteriol.* 134:316–329.

Guo, L.H., and R. Wu. 1983. Exonuclease III: Use for DNA sequence analysis and in specific deletions of nucelotides, 60–95 In R. Wu, L. Grossman, and K. Moldave, (ed.), Methods in Enzymology, vol. 100. Nucleic Acids. Academic, New York.

Hanahan, D., and M. Meselson. 1983. Plasmid screening at high colony density, 333–342. In R. Wu, L. Grossman, and K. Moldave (ed.), Methods in Enzymology, vol. 100. Recombinant DNA. Academic, New York.

Hansen, J.N. 1976. Electrophoresis of ribonucleic acid on a polyacrylamide gel which contains disulfide cross-linkages. *Anal. Biochem.* 76:37–44.

Hardies, S.C., R.K. Patient, R.D. Klein, F. Ho, W.S. Reznikoff, and R.D. Wells. 1979. Construction and mapping of recombinant plasmids used for the preparation of DNA fragments containing the *Escherichia coli* lactose operator and promoter. *J. Biol. Chem.* 254: 5527–5534.

Harris, J.E., K.F. Chater, C.J. Burton, and J.M. Piret. 1983. The restriction mapping of C gene deletions in Streptomyces bacteriophage 0C31 and their use in cloning vector development. *Gene* 22:167–174.

Hayashi, K. 1980. A cloning vehicle suitable for strand separation. *Gene* 11:109–115.

Heffron, F., R. Sublett, R.W. Hedges, A. Jacob, and S. Falkow. 1975. Origin of the TEM beta-lactamase found on plasmids. *J. Bacteriol.* 122:250–256.

Hindley, J. 1983. DNA sequencing. Elsevier, Amsterdam.

Hinnen, A., J.B.Hicks and G.R. Fink. 1978. Transformation of yeast. *Proc. Natl. Acad. Sci. U.S.A.* 75:1929–1933.

Hofschneider, P.H., and W. Goebel, (ed.). 1982. Gene cloning in organisms other than *E. coli.* Current Topics in Microbiol. Immunol., vol. 96. Springer-Verlag, New York.

Hohn, B., and J. Collins. 1980. A small cosmid for efficient cloning of large DNA fragments. *Gene* 11:291–298.

Hohn, B., and A. Hinnen. 1980. Cloning with cosmids in *E. coli* and yeast, 169–183. In J. Setlow and A. Hollaender, (ed.), Genetic Engineering: Principles and Methods. Plenum, New York.

Hollenberg, C.P. 1982. Cloning with 2-μm DNA vectors and the expression of foreign genes in *Saccharomyces cerevisiae. Curr. Top. Microbiol. Immunol.* 96:109–144.

Holmes, D.S., and M. Quigley. 1981. The rapid boiling method for the preparation of bacterial plasmids. *Anal. Biochem.* 114:193–197.

Hu, J.C., B.D. Coté, E. Lund, and J.E. Dahlberg. 1983. Isolation and characterization of genomic mouse DNA clones containing sequences homologous to tRNAs and SS rRNA. *Nucleic Acids Res.* 11:4809–4821.

Hutton, J.A. 1977. Renaturation kinetics and thermostability of DNA in aqueous solutions of formamide and urea. *Nucl. Acids Res.* 4:3537–3555.

Ikemura, T., and J.E. Dahlberg. 1973. Small RNA of *Escherichia coli.* I. characterization of polyacrylamide gel electrophoresis and fingerprinting. *J. Biol. Chem.* 248:5024–5032.

Jackson, D.A., R.H. Symons, and P. Berg. 1972. Biochemical method for inserting new genetic information into DNA of simian virus 40: circular SV40 DNA molecules containing lambda phage genes and the galactose operon of *Escherichia coli. Proc. Natl. Acad. Sci. U.S.A.* vol 69:2904–2909.

Jacobsen, H., H. Klenow, and K. Ovargaard-Hansen. 1974. the N-terminal amino-acid sequences of DNA polymerase I from *Escherichia coli* and of the large and the small fragments obtained by a limited proteolysis. *Eur. J. Biochem.* 45:623–627.

Jorgensen, J.H., J.C. Less, and G.A. Alexander. 1977. Rapid penicillinase paper strip test for detection of beta-lactamase-producing *Haemophilus influenzae* and *Neisseria gonorrhoeae. Antimicrob. Agents Chemother.* 11:1087–1088.

Jorgensen, R.A., D.E. Berg, B. Allet, and W.S. Reznikoff. 1979. Restriction enzyme cleavage map of Tn10, a transposon which encodes tetracycline resistance. *J. Bacteriol.* 137:681–685.

Kado, C.I., and S-T Liu. 1981. Rapid procedure for the detection and isolation of large and small plasmids. *J. Bacteriol.* 145:1365–1373.

Kafatos, F.C., C.W. Jones, and A. Efstratiadis. 1979. Determination of nucleic acid sequence homologies and relative concentrations by a dot hybridization procedure. *Nucl. Acids Res.* 7:1541–1552.

Karn, J., S. Brenner, L. Barnett, and G. Cesareni. 1980. Novel bacteriophage lambda cloning vector. *Proc. Natl. Acad. Sci. U.S.A.* 77:5172–5176.

Kingsman, A.J., L. Clarke, R.K. Mortimer, and J. Carbon. 1979. Replication in *Saccharomyces cerevisiae* of plasmid pBR313 carrying DNA from the yeast *trp*I region. *Gene* 7:141–152.

Kirby, K.S. 1968. Isolation of nucleic acids with phenolic solvents. In L. Grossman, and K. Moldave (ed.), Methods in Enzymology, vol. 12B:87–100.

Klee, H.J., F.F. White, V.N. Iyer, M.P. Gordon, and E.W. Nester. 1983. Mutational analysis of the virulence region of an *Agrobacterium tumefaciens* Ti-plasmid. *J. Bacteriol.* 153, 878–883.

Kornberg, A. 1974. DNA Synthesis. Freeman, San Francisco.

Kushner, S.R. 1978. An improved method for the transformation of *Escherichia coli* with ColE1-derived plasmids, 17–23. In H.B. Boyer, S. Nicosia (ed.), Genetic Engineering. Elsevier/North Holland, Amsterdam.

Langer-Safer, P.R., M. Levine, and D.C. Ward. 1982. Immunobiological method for mapping genes on *Drosophila* polytene chromosomes. *Proc. Natl. Acad. Sci. U.S.A.* 79:4381–4385.

Larsen, G.P., K. Itakura, H. Ito, and J.J. Rossi. 1983. *Saccharomyces cerevisiae* actin-*Escherichia coli lac*2 gene fusions: synthetic oligonucleotide-mediated deletion of the 309 base pair intervening sequence in the actin gene. *Gene* 22:31–39.

Laskey, R.A. 1980. The use of intensifying screens or organic scintillators for visualizing radioactive molecules resolved by gel electrophoresis, 363–371. In L. Grossman, and K. Moldave (ed.), Methods in Enzymology, vol. 65. Nucleic Acids I. Academic, New York.

Leary, J.J., D.J. Brigati, and D.C. Ward. 1983. Rapid and sensitive colorimetric method for visualizing biotin-labeled DNA probes hybridized to DNA or RNA immobilized on nitrocellulose: Bio-blots. *Proc. Natl. Acad. Sci. U.S.A.* 80:4045–4049.

Leder, P., D. Tremeier, and L. Enquist. 1977. EK2 derivatives of bacteriophage lambda useful in the cloning of DNA from higher organisms: the λgtWES system. *Science* 196:175–177.

Lederberg, E.M., and S.N. Cohen. 1974. Transformation of *Salmonella typhimurium* by plasmid deoxyribonucelic acid. *J. Bacteriol.* 119:1072–1074.

Lee, S.Y., V. Krsmanovic, and G. Brawerman. 1971a. Initiation of polysome formation in mouse sarcoma 180 ascites cells. Utilization of cytoplasmic messenger ribonucleic acid. *Biochem.* 10:895–900.

Lee, S.Y., J. Mendecki, and G. Brawerman. 1971b. A polynucleotide segment rich in adenylic acid in the rapidly-labeled polyribosomal RNA component of mouse sarcoma 180 ascites cells. *Proc. Natl. Acad. Sci. U.S.A.* 68:1331–1335.

Legerski, R.J., J.L. Hodnett, and H.B. Gray Jr. 1978. Extracellular nucleases of *Pseudomonas Bal*31 III. Use of the double-strand deoxyriboexonuclease activity as the basis of a convenient method for the mapping of fragments of DNA produced by cleavage with restriction enzymes. *Nucl. Acids Res.* 5:1445–1463.

Lehman, I.R., G.G. Roussos, and E.A. Pratt. 1962. Deoxyribonucleases of *Escherichia coli*. 2 purification of, properties of a ribonucleic acid-inhibitable endonuclease. *J. Biol. Chem.* 237:819–822.

Lehrach, H., D. Diamond, J.M. Wozney, and H. Boedtker. 1977. RNA molecular weight determination by gel electrophoresis under the denaturing condition, a critical reexamination. *Biochem.* 16:4743–4751.

Lewin, B. 1977. Phage lambda, 274–533. In B. Lewin (ed.), Gene Expression, vol. 3. Plasmids and Phages. Wiley, New York.

Levy, A., E. Frei, and M. Noll. 1980. Efficient transfer of highly resolved small DNA fragments from polyacrylamide gels to DBM paper. *Gene* 11:283–290.

Levy, S.B., B. Marshall, D. Rouse-Eagle, and A. Onderdonk. 1980. Survival of *Escherichia coli* host-vector systems in the mammalian intestine. *Science* 209:391–394.

Levy, S.B., and B. Marshall. 1981. Risk assessment studies of *E. coli* host-vector systems. *Recom. DNA Tech. Bull.* 4:91–98.

Lindberg, U., and T. Persson. 1982. Isolation of mRNA from KB-cells by affinity chromatography on polyuridylic acid covalently linked to sepharose. *Eur. J. Biochem.* 31:346–354.

Liu, C.P., P.W. Tucker, J.F. Mushinski, and F.R. Blattner. 1980. Mapping of heavy chain genes for mouse immunoglobulin M and D. *Science* 209:1348–1353.

Lobban, P.E., and A.D. Kaiser. 1973. Enzymatic end to end joining of DNA molecules. *J. Mol. Biol.* 78:453–471.

Loening, U.E. 1967. The fractionation of high-molecular-weight ribonucleic acid by polyacrylamide-gel electrophoresis. *Biochem. J.* 102:251–257.

Lorian, V. 1980. Antibiotics in Laboratory Medicine. Williams and Wilkins, Baltimore.

Loucks, E., G. Chaconas, R.W. Blackesley, R.D. Wells, and J.H. Van de Sande. 1979. Antibiotic-induced electrophoretic mobility shifts of DNA restriction fragments. *Nucl. Acids Res.* 6:1869–1879.

Lusky, M., and M. Botchan. 1981. Inhibition of SV40 replication in simian cells by specific pBR322 DNA sequences. *Nature* 293:79–81.

Maas, R. 1983. An improved colony hybridization method with significantly increased sensitivity for detection of single genes. *Plasmid* 10:296–298.

Macrina, F.L., J.A. Tobian, K.R. Jones, R.P. Evans, and D.B. Clewell. 1982. A cloning vector able to replicate in *Escherichia coli* and *Streptococcus sanguis*. *Gene* 19:345–353.

Mandel, M., and A. Higa. 1970. Calcium-dependent bacteriophage DNA infection. *J. Mol. Biol.* 53:159–162.

Maniatis, T., R.C. Hardison, E. Lacy, J. Lamer, C. O'Connell, D. Quan, G.K. Sim, and A. Efstratiadis. 1978. The isolation of structural genes from libraries of eukaryotic DNA. *Cell* 15:687–701.

Maniatis, T., E.F. Fritsch, and J. Sambrook. 1982. Molecular Cloning. A Laboratory Manual. Cold Spring Harbor Laboratory, Cold Spring Harbor, New York.

Martens, P.A., and D.A. Clayton. 1977. Strand breakage in solutions of DNA and ethidium bromide exposed to visible light. *Nucl. Acids Res.* 4:1393–1407.

Mathew, M., R.W. Hedges, and J.T. Smith. 1979. Types of β-lactamase determined by plasmids in gram-negative bacteria. *J. Bacteriol.* 138:657–662.

Maxam, A.M., and W. Gilbert. 1977. A new method for sequencing DNA. *Proc. Natl. Acad. Sci. U.S.A.* 74:560–564.

Maxam, A.M., and W. Gilbert. 1980. Sequencing end-labeled DNA with base-specific chemical cleavages, 499–560. In L. Grossman, and K. Moldave (ed.), Methods in Enzymology, vol. 65. Nucleic Acids. Academic, New York.

McKenney, K., H. Shimatake, D. Court, U. Schmeissener, C. Brady, and M. Rosenberg. 1981. A system to study promoter and terminator signals recognized by *E. coli* RNA polymerase, 383–415. In J. Chirikjean, and J. Papas (ed.), Gene Application and Analysis Vol II: Structural Analysis of Nucleic Acids. Elsevier/North Holland, New York.

McKnight, S.L., and E.R. Gavis. 1980. Expression of the herpes thymidine kinase gene in *Xenopus laevis* oocytes: an assay for the study of deletion mutants constructed in vitro. *Nucl. Acids Res.* 8:5931–5939.

Mercer, A.A., and J.S. Loutit. 1979. Transformation and transfection of *Pseudomonas aeruginosa*: effects of metal ions. *J. Bacteriol.* 140:37–42.

Mercola, K.E., and M.J. Cline. 1980. The potentials of inserting new genetic information. *New Engl. J. Med.* 303:1297–1300.

Merril, C.R., R.C. Switzer, and M.L. van Keuren. 1979. Trace polypeptides in cellular extracts and human body fluids detected by two-dimensional electrophoresis and a highly sensitive silver stain. *Proc. Natl. Acad. Sci. U.S.A.* 76:4335–4339.

Messing, J., B. Gronenborn, B. Muller-Hill, and P.H. Hofschneider. 1977. Filamentous coliphage M13 as a cloning vehicle: insertion of *Hind*II fragment of the *lac* regulatory region in M13 replicative form *in vitro*. *Proc. Natl. Acad. Sci. U.S.A.* 74:3642–3646.

Messing, J., R. Crea, and P.H. Seeburg. 1981. A system for shotgun DNA sequencing. *Nucl. Acids Res.* 9:309–321.

Messing, J., and J. Vieira. 1982. A new pair of M13 vectors for selecting either DNA strand of double-digest restriction fragments. *Gene* 19:269–276.

Meyers, J.H., D. Sanchez, L. P. Elwell, and S. Falkow. 1976. Simple agarose gel electrophoretic method for the identification and characterization of plasmid deoxyribonucleic acid. *J. Bacteriol.* 127:1529–1537.

Miller, J.H. 1972. Experiments in molecular genetics. Cold Spring Harbor Laboratory, Cold Spring Harbor, New York.

Miyazawa, Y., and C.A. Thomas, Jr. 1965. Nucleotide composition of short segments of DNA molecules. *J. Molec. Biol.* 11:223–237.

Modrich, P., and R. Rubin. 1979. Studies on *E. coli* DNA restriction and modification enzymes. *Fed. Proc.* 38:293.

Molloy, G., and L. Puckett. 1976. The metabolism of heterogeneous nuclear RNA and the formation of cytoplasmic messenger RNA in animal cells. *Prog. Biophys. Mol. Biol.* 31:1−38.

Morrison, D.A. 1977. Transformation in *Escherichia coli*: Cryogenic preservation of competent cells. *J. Bacteriol.* 132:349−351.

Moseley, S.L., I. Huq, A.R.M. Alim, M. So, M. Samadpour-Motalebi, and S. Falkow. 1980. Detection of enterotoxigenic *Escherichia coli* by DNA colony hybridization. *J. Infect. Dis.* 142:892−898.

Murray, N.E., W.J. Brammer, and K. Murray. 1977. Lamboid phages that simplify the recovery of *in vitro* recombinants. *Molec. Gen. Genet.* 150:53−61.

Nathans, D., and H.O Smith. 1975. Restriction endonucleases in the analysis and restructuring of DNA molecules. *Ann. Rev. Biochem.* 44:273−293.

Nichols, B.P., and C. Yanofsky. 1983. Plasmids containing the *trp* promoters of *Escherichia coli* and *Serratia marscesens* and their use in expressing cloned genes, 155−164. In R. Wu, L. Grossman, and K. Moldave (ed.), Methods in Enzymology, vol. 101. Recombinant DNA, Part C. Academic, New York.

Norgard, M.V., K. Keem, and J.J. Monahan. 1978. Factors affecting the transformation of *Escherichia coli* strain χ1776 by pBR322 plasmid DNA. *Gene* 3:279−292.

Novick, R.P. 1962. Micro-iodometeric assay for penicillinase. *Biochem. J.* 83:236−240.

O'Callaghan, C.H., A. Morris, S.M. Kirby, and A.H. Shingler. 1972. Novel method for detection of β-lactamase by using a chromogenic cephalosporin substrate. *Antimicrob. Agents Chemother.* 1:283−288.

O'Farrell, P.H., E. Kutter, and M. Nakanishi. 1980. A restriction map of the bacteriophage T4 genome. *Molec. Gen. Genet.* 179:421−435.

Okayama, H., and P. Berg. 1982. High-efficiency cloning of full-length cDNA. *Molec. Cell. Biol.* 2:161−170.

Olsen, R.H., G. DeBusschen, and W.R. McCombie. 1982. Development of broad host-range vectors and gene banks: self-cloning of the *Pseudomonas aeruginosa* PAO chromosome. *J. Bacteriol.* 150:60−69.

Osterlund, M., H. Luthman, S.V. Wilson, and G. Magnusson. 1982. Ethidium bromide-inhibited restriction endonucleases cleave one strand of circular DNA. *Gene* 20:121−125.

Parker, R.C., R.M. Watson, and J. Vinograd. 1977. Mapping of closed-circular DNA's by cleavage with restriction endonucleases and calibration by agarose gel electrophoresis. *Proc. Natl. Acad. Sci. U.S.A.* 74:851−855.

Peacock, A.C., and C.W. Dingman. 1968. Molecular weight estimation and separation of ribonucleic acid by electrophoresis in agarose-acrylamide composite gels. *Biochem.* 7:668−674.

Perret, C.J. 1954. Iodometric assay of pencillinase. *Nature* 174:1012−1013.

Petes, T.D., J.R. Broach, P.C. Wensink, L.M. Hereford, G.R. Fink, and D. Botstein. 1978. Isolation and analysis of recombinant DNA molecules containing yeast DNA. *Gene* 4:37−49.

Philippsen, P., R.A. Kramer, and R.W. Davis. 1978. Cloning of the yeast ribosomal DNA repeat unit in *Sst*I and *Hin*dIII lambda vectors using genetic and physical size selections. *J. Mol. Biol.* 123:371−386.

Porath, J., and P. Flodin. 1959. Gel filtration: A method for desalting and group separation. *Nature* 183:1657−1659.

Porath, J., J.-C. Janson, and T. Laas. 1971. Agar derivatives for chromatography, electrophoresis and gel-bound enzymes I. Desulphated and reduced cross-linked agar and agarose in spherical bead form. *J. Chromatogr.* 60:167−177.

Radloff, R., W. Bauer, and J. Vinograd. 1967. A dye-buoyant-density method for the detection and isolation of closed circular duplex DNA: the closed circular DNA in HeLa cells. *Proc. Natl. Acad. Sci. U.S.A.* 57:1514−152.

Razzaque, A., H. Mizusawa, and M.M. Seidman. 1983. Rearrangement and mutagenesis of a shuttle vector plasmid after passage in mammalian cells. *Proc. Natl. Acad. Sci. U.S.A.* 80:3010−3014.

Remault, E., H. Tsao, and W. Fiers. 1983. Improved plasmid vectors with a thermoinducible expression and temperature-regulated runaway replication. *Gene* 22:103−113.

Renz, H., and C. Kurz. 1984. A colorimetric method for DNA hybridization. Nucl. Acids Res. 12:3435−3444.

Rigby, T.W.J., M. Dieckmann, C. Rhodes, and P. Berg. 1977. Labeling deoxyribonucleic acid to high specific activity *in vitro* by nick translation with DNA polymerase I. *J. Molec. Biol.* 113:237−251.

Rimm, D.L., D. Horness, J. Kucera, and F.R. Blattner. 1980. Construction of coliphage lambda charon vectors with *Bam*HI cloning sites. *Gene* 12:301−309.

Roberts, R.J. 1976. Restriction endonucleases. *Crit. Rev. Biochem.* 4:123−164.

Roberts, R.J. 1982. Restriction and modification enzymes and their recognition sequences. *Nucl. Acids Res.* 10:r117−r144.

Roberts, T.M., R. Kacich, and M. Ptashne. 1979. A general method for maximizing the expression of a cloned gene. *Proc. Natl. Acad. Sci. U.S.A.* 76:760−764.

Rosen, C.G., I. Fedorcsak. 1966. Studies on the action of diethyl pyrocarbonate on proteins. *Biochem. Biophys. Acta.* 130:401−405.

Rosenberg, M., Y.-S. Ho, and A. Shatzman. 1983. The use of pKC30 and its derivatives for controlled expression of genes, 123−138. In R. Wu, L. Grossman, and K. Moldave (ed.), Methods in Enzymology, vol. 101. Recombinant DNA. Academic, New York.

Ross, G.W., and C.H. O'Callaghan. 1975. β-lactamase assays, 69−85. In J.H. Hash (ed.), Methods in Enzymology, vol. 43. Academic, New York.

Roychaudhury, R., and H. Kossel. 1971. Synthetic polynucleotides, enzyme synthesis of ribonucleotide terminated oligodeoxynucleotides and their use as primers for the enzyme synthesis of polydeoxynucleotides. *Eur. J. Biochem.* 22:310−320.

Roychaudhury, R., and R. Wu. 1980. Terminal transferase-catalysed addition of nucleotides to the 3′ termini of DNA, 43−62. In L. Grossman, and K. Moldave (ed.), Methods in Enzymology, vol. 65. Nucleic Acids. Academic, New York.

Rushizky, G.W., A. Greco, R. Hartly Jr., and H. Sober. 1963. Concentration and desalting of ribonucleases. *Biochem. Biophys. Res. Commun.* 16:311−314.

Sanger, F., S. Nicklen, and A.R. Coulsen. 1977. DNA sequencing with chain-terminating inhibitors. *Proc. Natl. Acad. Sci. U.S.A.* 74:5463−5467.

Sarver, N.J., C. Byrne, and P.M. Howley. 1982. Transformation and replication in mouse cells of a bovine papillomavirus-pML2 plasmid vector that can be rescued in bacteria. *Proc. Natl. Acad. Sci. U.S.A.* 79:7147−7151.

Schleif, R.F., and P.C. Wensink. 1981. Practical Methods in Molecular Biology. Springer-Verlag, New York.

Sealey, P.G., and E.M. Southern. 1982. Gel electrophoresis of DNA, 39−75. In D. Rickwood, and B.D. Hames (ed.), Gel Electrophoresis of Nucleic Acids: a practical approach. IRL, Oxford.

Seidman, G., M.H. Edgell, and P. Leclen. 1978. Immunoglobulin light chain structural gene sequences cloned in a bacterial plasmid. *Nature* 271:582−585.

Sharp, R.A., B. Sugden, and J. Sambrook. 1973. Detection of two restriction endonuclease activities in *Haemophilus parainfluenzae* using analytical agarose-ethidium bromide electrophoresis. *Biochem.* 12:3055−3063.

Shimatake, H., and M. Rosenberg. 1981. Purified λ regulating protein CII positively activates promoter for lysogenic development. *Nature* 292:128−132.

Shine, J., and L. Dalgarno. 1975. Determinant of cistron specifity in bacterial ribosomes. *Nature* 254:34−38.

Singer, B., H. Fraenkel-Conrat, and A. Tsugita. 1961. Purification of viral RNA by means of bentonite. *Virology.* 14:54−58.

Smith, G.E., and M.D. Summers. 1980. The bidirectional transfer of DNA and RNA to nitrocellulose or diazobenzylmethyl-paper. *Anal. Biochem.* 109:123.

Smith, H.O., and D. Nathans. 1973. A suggested nomenclature for bacterial host modification and restriction systems and their enzymes. *J. Mol. Biol.* 81:419–423.

Smith, H.O., and M.L. Bernstiel. 1976. A simple method for DNA restriction site mapping. *Nucl. Acids Res.* 3:2387–2398.

Smith, H.O., and D.B. Danner. 1981. Genetic transformation. *Ann. Rev. Biochem.* 50: 41–68.

Smith, S.S., and C.A. Thomas Jr. 1981. The two-dimensional restriction analysis of *Drosophila* DNAs: males and females. *Gene* 13:395–408.

Southern, E.M. 1975. Detection of specific sequences among DNA fragments separated by gel electrophoresis. *J. Mol. Biol.* 98:503–517.

Southern, E.M. 1979. Gel electrophoresis of restriction fragments, 152–182. In R. Wu (ed.), Methods in Enzymology, vol. 68. Recombinant DNA. Academic, New York.

Struhl, K., D.T. Stinchcomb, S. Scherer, and R.W. Davis. 1979. High frequency transformation of yeast: autonomous replication of hybrid DNA molecules. *Proc. Natl. Acad. Sci. U.S.A.* 76:1035–1039.

Studier, F. 1969. Effects of the conformation of single-stranded DNA on renaturation and aggregation. *J. Mol. Biol.* 41:199–209.

Studier, F.W. 1973. Analysis of bacteriophage T7 early RNAs and proteins on slab gels. *J. Mol. Biol.* 79:237–248.

Sutcliffe, J.G. 1979. Complete nucleotide sequence of the *Escherichia coli* plasmid pBR322. Cold Spring Harbor Symposium 43:77–90.

Swan, D., H. Aviv, and P. Leder. 1972. Purification and properties of biologically active messenger RNA for a myeloma light chain. *Proc. Natl. Acad. Sci. U.S.A.* 69:1967–1971.

Sykes, R.B., and M. Matthew. 1976. The β-lactamases of gram-negative bacteria and their role in resistance to β-lactam antibiotics. *J. Antimicrob. Chemother.* 2:115–157.

Tabak, H.F., and R.A. Flavell. 1978. A method for the recovery of DNA from agarose gels. *Nucl. Acids Res.* 5:2321–2332.

Talmadge, K., S. Stahl, and W. Gilbert. 1980. Eucaryotic signal sequence transports insulin antigen in *E. coli*. *Proc. Natl. Acad. Sci. U.S.A.* 77:3369–3373.

Tautz, D., and M. Renz. 1983. An optimized freeze-squeeze method for the recovery of DNA fragments from agarose gels. *Anal. Biochem.* 132:14–19.

Taylor, J.M. 1979. The isolation of eucaryotic messenger RNA. *Ann. Rev. Biochem.* 48:681–717.

Temin, H.M., and S. Mizutani. 1970. RNA-dependent DNA polymerase in virions of Rous sarcoma virus. *Nature* 226:1211–1213.

Tenover, F.C., L.W. Mayer, and F.E. Young. 1980. Physical map of the conjugal plasmid of *Neisseria gonorrhoeae*. *Infect. Immun.* 29:181–185.

Thomas, P.S. 1980. Hybridization of denatured RNA and small DNA fragments transferred to nitrocellulose. *Proc. Natl. Acad. Sci. U.S.A.* 77:5201–5205.

Thomas, P.S. 1983. Hybridization of denatured RNA transferred or dotted to nitrocellulose paper, 255–266. In R. Wu, L. Grossman, K. Moldave (ed.), Methods in Enzymology, vol. 100. Recombinant DNA. Academic, New York.

Thompson, J.A., R.W. Blakesley, K. Doran, C.J. Hough, and K.D. Wells. 1983. Purification of nucleic acids by RPC-5 analogue chromatography: peristaltic and gravity flow application., 368–399. In R. Wu, L. Grossman, and K. Moldave (ed.), Methods in Enzymology, vol. 100. Recombinant DNA. Academic, New York.

Thuring, R.W.J., J. Sanders, and P. Borst. 1975. A freeze-squeeze method for recovering long DNA from agarose gels. *Anal. Biochem.* 66:213–220.

Tiemeir, D., L. Enquist, and P. Leder. 1976. An improved derivative of a bacteriophage lambda EK2 vector useful in the cloning of recombinant molecules: λgtWESλB. *Nature* 263:526–527.

Tu, C.P.D., and S.N. Cohen. 1980. 3′-end labeling of DNA with [α ^{32}P] cordycepin-5′-triphosphate. *Gene* 10:177−183.

Ulrich, A., J. Shine, J. Chirgwin, R. Pictet, E. Tischer, W.J. Rutter, and H.M. Goodman. 1977. Rat insulin genes: construction of plasmids containing the coding sequences. *Science* 196:1313−1319.

Uriel, J., and J. Berges. 1966. Un nouveau support pour des séparations electrophorétiques: le gel mixte d'acrylamide-agarose. Acad. Sci. Comptes. Rendus. Part. C. 262:164−167.

Van de Hondel, C., W. Keegstra, W. Borrias, and G. Van Arkel. 1979. Homology of plasmids in strains of unicellular cyanobacteria. *Plasmid* 2:323−333.

Van Wezenbeek, P.M.G.F., T.J.M. Hulsebos, and J.G.G. Shoemakers. 1980. Nucleotide sequence of filamentous bacteriophage M13 DNA genome: comparison with phage Fd. *Gene* 11:129−148.

Verma, J.M. 1977. The reverse transcriptase. *Biochem. Biophys. Acta.* 473:1−38.

Vieira, J., and J. Messing. 1982. The pUC plasmids, a M13mp7 derived system for insertion mutagenesis and sequencing with synthetic universal primers. *Gene* 19:259−268.

Villa-Komaroff, L., A. Efstratiadis, S. Broome, P. Lomedico, R. Tizard, S.P. Naber, W.L. Chick, and W. Gilbert. 1978. A bacterial clone synthesizing proinsulin. *Proc. Natl. Acad. Sci. U.S.A.* 75:3727−3731.

Vincent, W.S., and E.S. Goldstein. 1981. Rapid preparation of covalently closed circular DNA by acridine yellow affinity chromatography. *Anal. Biochem.* 110:123−127.

Vogelstein, B., and D. Gillespie. 1979. Preparative and analytical purification of DNA from agarose. *Proc. Natl. Acad. Sci. U.S.A.* 76:615−619.

Wahl, G.M., M. Stern, and G.R. Stark. 1979. Efficient transfer of large DNA fragments from agarose gels to diazobenyloxymethyl paper and rapid hybridization by using dextran sulfate. *Proc. Natl. Acad. Sci. U.S.A.* 76:3683−3687.

Weislander, L. 1979. A simple method to recover intact high molecular weight RNA and DNA after electrophoretic separation in low-gelling temperature agarose gels. *Anal. Biochem.* 98:305−309.

White, F.F., H.J. Klee, and E.W. Nester. 1983. *In vivo* packaging of cosmids in transposon-mediated mutagenesis. *J. Bacteriol.* 153:1075−1078.

Williams, K.W. 1972. Solute-gel interactions in gel filtration. *Lab. Practice* 21:667−670.

Williams, B.L., and K. Wilson. 1975. A biologist's guide to principles and techniques of practical biochemistry. Elsevier, New York.

Williams, W.G., and F.R. Blattner. 1980. Bacteriophage lambda vectors for DNA cloning, 201−281. In J. Setlow, and A. Hollaender (ed)., Genetic Engineering: Principles and Methods, vol. 2. Plenum, New York.

Wilson, J.T., L.B. Wilson, J.K. de Riel, L. Villa-Komaroff, A. Efstratiadis, B.G. Forget, and S.M. Weissman. 1978. Insertion of synthetic copies of human globin genes into bacterial plasmids. Nucl. Acids Res. 5:563−581.

Winberg, G., and M.A. Hammarskjold. 1980. Isolation of DNA from agarose gels using DEAE paper. Application of restriction site mapping of adenovirus type 16 DNA. *Nucl. Acids Res.* 8:253−264.

Yamamoto, K.R., B.M. Alberts, R. Benzinger, L. Lawthorne, and G. Treiber. 1970. Rapid bacteriophage sedimentation in the presence of polyethylene glycol and its application to large-scale virus purification. *Virol.* 40:734−744.

Yang, R., J. Lis, and R. Wu. 1979. Elution of DNA from agarose gels after electrophoresis, 176−182. In R. Wu (ed.), Methods in Enzymology, vol. 68. Academic, New York.

Yoneda, Y., S. Graham, and F.E. Young. 1979. Restriction-fragment map of the template *Bacillus subtilis* bacteriophage SPO2. *Gene* 7:51−68.

Yuan, R. 1981. Structure and mechanism of multifunctional restriction endonucleases. *Ann. Rev. Biochem.* 50:285−315.

Zandvliet, G.M., and H.S. Jansz. 1976. Characterization and replication control of plasmids from *Enterobacter cloacae* DF13. *Eur. J. Biochem.* 62:439–449.

Zinder, N.D., and J.D. Boeke. 1982. The filamentous phage (Ff) as vectors for recombinant DNA—a review. *Gene* 19:1–10.

Zöllner, N., and J. Fellig. 1953. Nature of inhibition of ribonuclease by heparin. *Am. J. Physiology.* 173:223–228.

chapter 2

GENE CLONING IN *SACCHAROMYCES CEREVISIAE*

A.A. Potter, A. Nasim, R.S. Zitomer,
and C.P.Hollenberg

The yeast *Saccharomyces cerevisiae* has been used extensively for genetic studies, mainly due to its ease of manipulation in the laboratory and the availability of a wide range of mutations affecting various cellular processes (Hawthorne and Mortimer, 1978; Strathern *et al.*, 1981, 1982). Recent reports have shown the usefulness of yeast as a host for the study of relevant topics in the field of molecular biology. For example, Lörincz and Reed (1984) have shown that substantial sequence homology exists between a yeast cell division gene and a vertebrate oncogene, thus allowing the analysis of oncogene function under haploid conditions.

Most *S. cerevisiae* strains carry an autonomously replicating extrachromosomal element analogous to a plasmid in bacteria, called 2-μm DNA. This element is present at approximately 50−100 copies per cell and is maintained in a stable manner, thus making it potentially useful as a cloning vector. That, combined with the development of an efficient transformation system (Hinnen *et al.*, 1978; Beggs, 1978; Struhl *et al.*, 1979) makes *S. cerevisiae* a popular and useful host for gene cloning experiments. As a eukaryote, *S. cerevisiae* could, in certain cases, be a more suitable host than bacteria for the cloning of genes from yeast and other eukaryotes, especially if foreign or controlled gene expression is desired. This is illustrated by a recent report describing the production of a hepatitis B vaccine from recombinant yeast (McAleer *et al.*, 1984). Also, new cloning vectors that control not only gene expression, but also secretion of the gene product are being developed (Skipper *et al.*, 1984).

This chapter describes methods for gene cloning and expression in yeast, including various procedures used for genetic characterization of potential recombinant strains. For illustrative purposes, we describe the cloning of the *LEU2* gene of *S. cerevisiae*. For recent reviews on yeast transformation, see Hinnen and Meyhack (1982) and Hollenberg (1982). In addition, several recent papers describe useful methods that will not be discussed here (Murray and Szostak, 1983; Crosby and Thomas, 1983; Struhl, 1983; Wu *et al.*, 1983).

I. DNA Isolation Techniques

METHOD 1:   TOTAL YEAST DNA EXTRACTION

PROCEDURE	COMMENTS
1. Grow cells of a Leu$^+$ strain of *S. cerevisiae* to late exponential phase in 5.0 mL of YEPD broth. Harvest by centrifugation and resuspend the cells in 1.0 mL of 1.0M sorbitol.	**1. YEPD Broth:** 1% yeast extract 2% peptone 2% dextrose Yeast cells lyse more readily following growth in media containing galactose, which can be substituted for dextrose in YEPD.
2. Transfer the cells to a 1.5 mL microfuge tube, pellet by spinning for 30–60 seconds in a microfuge, and resuspend in 0.5 mL of 1.0M sorbitol, 50 mM potassium phosphate (pH 7.5), 14 mM β-mercaptoethanol.	**2.** Add the β-mercaptoethanol to the sterilized solution immediately prior to use. Treatment of cells with reducing agents such as mercaptoethanol leads to more efficient **PROTOPLAST FORMATION**.
3. Add 25–50 units of Glusulase, Helicase or 500 μg Zymolyase and incubate at 30°C, 30 minutes.	**3.** Glusulase, a crude mixture of glucuronidase and sulfatase, can be obtained from Endo Laboratories, Garden City, New York. Zymolyase (Miles Laboratories, Rexdale, Ontario), Mutanase (Novo, Bagsvaerd, Denmark) and Helicase (Société Chimique Pointet-Girard, Villeneuve-la-Garenne, France) are quite useful substitutes for Glusulase. Protoplast formation may be monitored by placing a drop of cells in a 2% solution of SDS and comparing it with the density of a similar dilution in sorbitol. The concentration of the enzyme required varies with the strains used.
4. Spin the protoplasts and resuspend in 0.5 mL of 50 mM EDTA, pH 8.5. Add 20 μL of SDS (10%) and 1 μL of diethyloxydiformate (in a fume hood).	**4.** The combination of EDTA and diethyloxydiformate (DED) prevents endogenous nuclease digestion of the DNA. If DED is used, the DNA sometimes cannot be used for cloning, only for restriction analysis, transformation, etc.
5. Heat to 70°C for 15 minutes in the fume hood.	
6. Chill on ice and add 100 μL of chilled 5M potassium acetate solution, mix gently, and leave on ice for 60 minutes.	**6. Potassium Acetate Solution:** 3M potassium acetate 2M glacial acetic acid

PROCEDURE	COMMENTS

A fine white precipitate, consisting of protein and some DNA will form.

7. Spin for 10 minutes in a microfuge (13,000 × g), decant the supernatant to a new microfuge tube and fill with ethanol at room temperature. Mix gently.

7. Be careful not to remove any of the white precipitate with the supernatant. The DNA should be visible as a fibrous precipitate upon addition of the ethanol.

8. Spin up to 10 minutes in a microfuge and drain. Briefly dry under vacuum or at 37°C for 15 minutes.

9. Dissolve the DNA pellet in 50 μL of 10 mM Tris-HCl, pH 7.5, 1 mM EDTA, containing 10 μg/mL of boiled RNase A.

9. The RNase solution should be placed in a boiling water bath for 10 minutes prior to use in order to inactivate any DNase's which may be present.

10. After 3 hours at room temperature, remove residual protein by 3 successive phenol-chloroform (1:1) extractions (or until the solution is clear).

10. See Chapter 1, Section I (p. 3) for a detailed description of the **PHENOL EXTRACTION PROCEDURE**.

11. Precipitate the DNA with ethanol (Chapter 1, Section I, p. 4), spin 10 minutes in a microfuge, dry and dissolve the DNA in 50 μL of 20 mM Tris-HCl, pH 7.5, 1 mM EDTA.

11. This procedure is basically that of Struhl *et al.* (1979), and should yield 1−2 μg of DNA per mL of culture used. The DNA is of sufficient purity to be cut with restriction endonucleases and used for gel analysis or for transformation. If DNA of very high purity is required, the method of Cryer *et al.* (1975) should be used.

METHOD 2: ISOLATION OF COVALENTLY-CLOSED-CIRCULAR DNA FROM *S. CEREVISIAE*

Recently, a procedure was developed for the isolation of covalently-closed-circular (CCC) DNA from *S. cerevisiae* (Devenish and Newlon, 1982). The method is rapid, and as with procedures for the isolation of plasmid DNA from bacteria, relies upon alkaline denaturation of linear DNA to enrich for circular DNA.

PROCEDURE	COMMENTS

1. Grow cells in supplemented minimal medium (40 mL) to a density of approximately 5×10^7 cells/mL. Harvest by centrifugation.

1. In the case of YRp7 *LEU2* recombinant plasmids, cells should be grown in minimal medium supplemented with histidine (for *S. cerevisiae* host strain YF135 or LL20).

2. Wash once with an equal volume of distilled water, harvest by centrifugation, and resuspend the pellet in 20 mL of pretreatment buffer. Incubate at room temperature for 10 minutes.

2. Pretreatment Buffer:

0.2M Tris-HCl
1.2M sorbitol
0.1M EDTA
0.1M β-mercaptoethanol, pH 9.1

Steps 2−4 are concerned with **PROTOPLAST PREPARATION** and the procedure for *S. CEREVISIAE* **TRANSFORMATION** can likely be substituted (Section III, p. 137).

PROCEDURE

3. Wash the cells twice with 25 mL of SCE buffer and resuspend in the same at a cell density of no greater than 1×10^8/mL.

4. Add Zymolyase-60000 to a concentration of 100 µg/mL and incubate at 37°C for 25 minutes. Harvest by gentle centrifugation.

5. Carefully resuspend the pellet in 0.5 mL of 25% sucrose, 50 mM Tris-HCl, pH 8.0. Stir for 90 seconds at 100 rpm with a teflon-coated magnetic bar while adding 9.5 mL of lysis buffer in drop-wise fashion.

6. Incubate at 37°C for 25 minutes and then adjust the pH to 8.5–8.9 by adding 2M Tris-HCl, pH 7.0. Stir with a magnetic bar while adding the Tris-HCl, and monitor the pH with paper strips.

7. Add solid NaCl to a concentration of 3% (w/v) and incubate at room temperature for 30 minutes.

8. Add an equal volume of phenol saturated with 3% (w/v) NaCl, mix by stirring fo 10 seconds at approximately 300 rpm and a further 2 minutes at 100 rpm. Separate the phases by centrifugation and recover the aqueous phase.

9. Add sodium acetate to 0.3M and precipitate the nucleic acids by the addition of 2 volumes of ethanol. Leave at -20°C for at least 2 hours. Collect the DNA by centrifugation for 25 minutes at $11,200 \times g$, and dissolve the pellet in 10 mM Tris-HCl, pH 8.0, 1 mM EDTA.

COMMENTS

3. SCE Buffer:

1.0M sorbitol
0.1M sodium citrate
60 mM EDTA, pH 5.8

4. Glusulase can be substituted for Zymolyase if desired.

5. Lysis Buffer:

50 mM Tris-HCl
0.02M EDTA
1% (w/v) SDS
Adjusted at 23°C to pH 12.45

As with other alkaline lysis procedures for isolating plasmid DNA, the pH of the lysis buffer is critical. When the pH is greater than 12.5 some plasmids are irreversibly denatured. At lower pH (≤ 12.2), inadequate denaturation of the linear DNA will occur.

6. Once the lysate is neutralized, the viscosity should decrease dramatically. If this does not occur, it is probably due to a faulty lysis buffer.

8. PHENOL EXTRACTION in the presence of 3% (w/v) NaCl removes single-stranded DNA from the aqueous phase. Thus, circular DNA is enriched at this step.

9. The DNA recovered following ETHANOL PRECIPITATION is suitable for analysis by agarose gel electrophoresis, restriction endonuclease digestion, or can be used to transform *E. coli*.

METHOD 3: EXTRACTION OF AMPLIFIABLE PLASMID DNA FROM *E. COLI*

PROCEDURE

1. Dilute a fresh overnight culture of *E. coli* C600/YRp7 into 100 mL–1L of minimal medium supplemented with 0.1% casamino acids. Incubate with shaking at 37°C until the cell density reaches about 3×10^8 cells/mL.

COMMENTS

1. It is useful to include an antibiotic to which the plasmid specifies resistance (ampicillin in the case of YRp7) in the growth medium. This will ensure that each cell retains the plasmid.

2. For amplification, add chloramphenicol to a final concentration of 200 μg/mL and continue incubation for 16–24 hours.

2. If the plasmid specifies resistance to chloramphenicol, kanamycin or spectinomycin may be used. If the plasmid is non-amplifiable, omit this step and grow cells to stationary phase.

3. Harvest the cells by centrifugation and resuspend in 4.0 mL of 25 mM Tris-HCl, pH 8.0, 5 mM EDTA. Incubate at room temperature for 5 minutes.

4. Add 8.0 mL of 1% SDS, 0.2 N NaOH, mix gently by inversion, and incubate 5 minutes on ice.

4. The SDS solution should be made weekly, using a 10M NaOH stock solution. The cells should lyse immediately upon addition of the SDS solution.

5. Add 6.0 mL of 5M potassium acetate solution (pH 4.8), mix gently, and leave on ice for 5 minutes.

5. The potassium acetate solution consists of 3M potassium acetate and 2M glacial acetic acid. A white precipitate will form immediately upon addition of this solution. The purpose of this step is to remove single-stranded chromosomal DNA and protein.

6. Spin for 10 minutes at 10,000 rpm in a Sorval SS34 rotor at 4°C, and carefully decant the supernatant into a clean centrifuge tube.

6. Be careful not to include any of the white pellet while decanting the supernatant.

7. Add 2.5 mL of 5 M NaCl, 5.0 mL of 50% polyethylene glycol 6000, and incubate at 4°C for at least 5 hours (overnight if convenient).

7. All of the polyethylene glycol should be drained from the tube following centrifugation. Resuspend the pellet gently.

8. Harvest the precipitate by spinning at 5,000 rpm for 5 minutes in a Sorval SS34 rotor. Resuspend in 5.0 mL of TE buffer.

9. Add 5.0 g of solid CsCl, 0.35 mL of ethidium bromide (10 mg/mL), and spin 48 hours at 45,000 rpm in the type 50Ti rotor of a Beckman ultracentrifuge.

10. View the tube under ultraviolet illumination and remove the thick lower band with a syringe. Remove the ethidium bromide by repeated extraction with CsCl-saturated isopropanol. Add 2 volumes of distilled water and precipitate the DNA with ethanol (63–67% final concentration). Leave overnight at −20°C.

10. Following centrifugation, two bands should be visible when viewed under ultraviolet illumination, the upper containing linear DNA and the bottom containing plasmid DNA. The isopropanol extractions should be carried out until all traces of the ethidium bromide are gone. The addition of distilled water prior to ETHANOL PRECIPITATION prevents the CsCl from precipitating along with the DNA.

11. Harvest the DNA by spinning at 12,000 rpm, for 15 minutes in a Sorval HB-4 rotor. Wash the pellet with 70% ethanol and suspend the final DNA in 100–500 μL of TE.

11. Do not over-dry the DNA pellet prior to dissolving in TE. The concentration of the DNA can be determined as described in Chapter 1, Section XVII-M; p. 102.

II. Cloning in Yeast

A. CLONING VECTORS

Although a number of different *S. cerevisiae* cloning vectors have been described, they all have common features. They are composed of at least two components, one of bacterial DNA and the other of yeast DNA. The bacterial DNA is that of a plasmid such as pBR322, so that the vector will replicate in *E. coli*, while the yeast DNA contains at least one gene which can be used as a selectable marker in *S. cerevisiae*, illustrated in Figure 2.1. A recent report (Singh *et al.*, 1982) described the transformation of yeast with single-stranded DNA. ssDNA vectors transformed yeast cells 10–30 times more efficiently than dsDNA of the same sequence.

B. INTEGRATING VECTORS

Yeast integrating plasmids (YIp) consist of a suitable bacterial replicon plus a gene from *S. cerevisiae*, which can be used as a selective marker during transformation (Hinnen *et al.*, 1978). The yeast DNA also serves as a portable region of homology that promotes recombination of the plasmid with the nuclear DNA of the recipient strain. Although very little is known about the recombination process in *S. cerevisiae*, Hicks *et al.* (1979) observed that cutting the plasmid with a restriction endonuclease that recognizes a single site within the yeast gene increased the transformation efficiency. This suggests that efficient integration of DNA into the chromosome requires free ends on the DNA molecule (see also Orr-Weaver *et al.*, 1981).

Generally, YIp plasmids exhibit low transformation efficiency and yield low copy number (usually one) in the transformed recipient cell (Table 2.1). However, since integration of the plasmid is accomplished, YIp vectors have some notable applications in the analysis of the yeast genome.

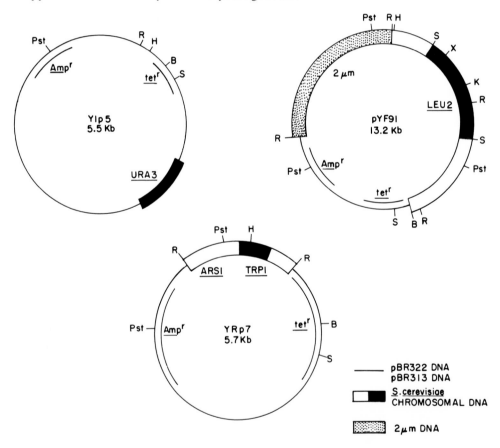

FIGURE 2.1 Physical maps of an integration vector, YIp5 (Struhl *et al.*, 1979), a 2-µm DNA vector, pYF91 (Storms *et al.*, 1979) and an autonomously replicating vector, YRp7 (Stinchcomb *et al.*, 1979). The relative positions of the ampicillin resistance *(amp^r)*, tetracycline resistance (*tet^r*), *LEU2*, *TRP1* and *URA3* genes are shown. Restriction endonuclease cleavage sites are: B, *Bam*HI; H, *Hind*III; K, *Kpn*I; Pst, *Pst*I; R, *Eco*RI; S, *Sal*I; X, *Xho*I.

TABLE 2.1 Properties of Bacterial–Yeast Hybrid Cloning Vectors

Vector Type	Transformation Frequency (per Viable Spheroplast)	Transformation Efficiency (per μg DNA)	Stability in Nonselection Media (Percent Loss per Generation)
Yeast-integrating plasmid (YIp)	10^{-7}	$1-10$	0.05
Yeast-replicating plasmid (YRp)	10^{-4}	10^2-10^4	$5-10$
Yeast episomal plasmid (YEp; 2-μm vector)	10^{-3}	10^3-10^5	$1-3$
Centromeric plasmid	10^{-4}	10^2-10^4	0.05

If one digests the DNA of transformed yeast cells with the restriction endonuclease originally used to cut within the yeast gene, the integrated cloning vector can be recovered (Figure 2.2). If one cuts with another suitable restriction endonuclease, it is possible to recover a mutated chromosomal gene, rather than its allele present on the vector molecule. Thus, it is possible to carry out detailed analyses (DNA sequencing etc.) of mutant alleles using this procedure (see Roeder and Fink, 1980).

Since yeast cells transformed with an integration vector contain two copies of the gene present on the vector, as is shown in Fig. 2.2, it is possible to introduce genes into yeast which have been mutagenized *in vitro* at specific sites. This work was originally carried out by Scherer and Davis (1979) and is illustrated in Figure 2.3. In this experiment, a specific deletion was introduced into the *HIS3* region of a *HIS3 URA3* plasmid, which was subsequently used to transform a *ura3 HIS3* strain of *S. cerevisiae* to a Ura$^+$ phenotype. Since the *URA3* locus is smaller than *HIS3* (Scherer and Davis, 1979), integration of the plasmid occurred primarily at *HIS3*. By growing purified transformants under nonselective conditions, approximately 1% will spontaneously lose the vector (see Table 2.1) and become Ura$^-$. Some of the these Ura$^-$ cells will have lost the chromosomal *HIS3* gene rather than the deleted version originally present on the vector plasmid, due to aberrant excision, and these cells will be phenotypically Ura$^-$His$^-$. Thus, the effect of a specific deletion generated *in vitro* can be studied *in vivo*.

C. AUTONOMOUSLY REPLICATING VECTORS

Plasmids containing *ARS* sequences (autonomous replication sequences) contain a bacterial replicon, a yeast gene for selection and an origin of replication, *ARS*, which can function in yeast (Struhl *et al.*, 1979; Kingsman *et al.*, 1979). These sequences occur frequently, about 25 times per chromosome. Similar sequences have also been recovered from mitochondrial DNA of *S. cerevisiae* (Hyman *et al.*, 1982). Autonomously replicating vector molecules (YRp) transform *S. cerevisiae* at a relatively high frequency (Table 1) and exist as low copy number plasmids within the nucleus of the cell. Yeast cells transformed with YRp7 (Fig. 1) are unstable for the plasmid-associated *TRP$^+$* marker (Table 1) and the plasmid segregates in unstable fashion at mitotic and meiotic divisions. Kingsman *et al.* (1979) demonstrated that when diploid cells carrying YRp7 are sporulated, all products of the meiotic segregation are Trp$^-$.

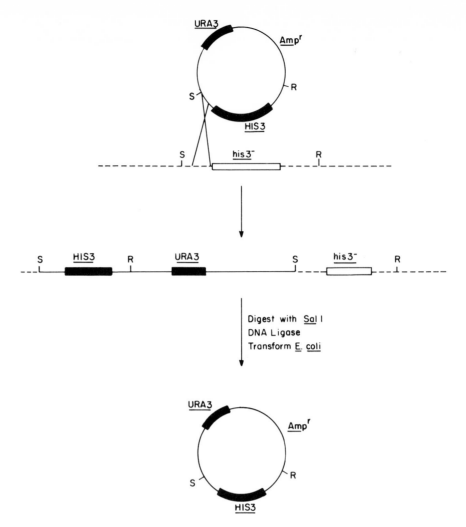

FIGURE 2.2 Integrative recombination of a Y1p5::*HIS3* plasmid with the yeast chromosome XV and its subsequent recovery. In order to recover plasmid DNA from such a transformant, total DNA is digested with *Sal*I, ligated and then used to transform *E. coli*. Abbreviations are the same as Figure 2.1.

————, bacterial DNA; _ _ _ _ _, yeast chromosomal DNA; ▬▭, yeast genes.

D. 2-μm DNA VECTORS

Most strains of *S. cerevisiae* contain a small circular DNA molecule approximately 2 micrometers in length, at a copy number of 50−100 molecules per cell. Its meiotic segregation pattern, $4^+:0^-$ (Livingston, 1977), is typical of cytoplasmic determinants. However, its nucleosome structure and replication suggest a close association with the nuclear DNA (Seligy *et al.*, 1980). The genetic and physical properties of a 2-μm DNA have been recently reviewed elsewhere (Broach, 1982; Guerineau, 1979).

Cloning vectors which contain a bacterial replicon, a yeast gene, and all or part of the 2-μm plasmid are able to transform *S. cerevisiae* cells with high efficiency and relatively high stability (Table 2.1). Such vectors are maintained as plasmids in *S. cerevisiae* and segregate in a non-Mendelian fashion. Therefore, plasmid DNA can be recovered from transformed yeast cells by isolating total DNA and using this

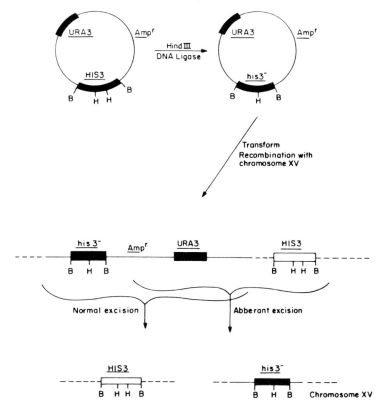

FIGURE 2.3 Replacement of the *HIS3* gene with a *his3* deletion derivative constructed by *in vitro* mutagenesis. See the text for details.

————, bacterial DNA; _ _ _ _ _, yeast chromosomal DNA; ▮▯, yeast genes. Based on the work of Scherer and Davis (1979). Abbreviations are the same as in Figure 2.1.

mixture to transform *E. coli*, selecting for the marker present on the bacterial replicon moiety of the vector (ampicillin-resistance, for example). It is possible to isolate 10–20 bacterial transformants per microgram of yeast DNA used (Broach *et al.*, 1979).

A 2-μm DNA vector containing the *E. coli* β-lactamase *(bla)* gene under the control of the yeast *ADH1* promoter was recently constructed (Reipen *et al.*, 1982). This vector is useful, in that it can be used to transform wild or industrial yeast strains which do not carry an auxotrophic mutation for selection. Transformants cannot be selected but can be identified by a β-lactamase plate assay (Chapter 1, Section XVII-R, p. 110).

E. CENTROMERE VECTORS

Bacterial-yeast hybrid plasmids which contain centrometric DNA are relatively stable in yeast and segregate in a similar fashion as normal chromosomes. Clarke and Carbon (1980) cloned the *CEN3* region of *S. cerevisiae* and showed that the cloned region was able to stabilize (in *cis*) plasmids carrying an *ARS1* sequence. Likewise, *CEN11* (Hsiao and Carbon, 1981) and *CEN4* (Stinchcomb *et al.*, 1982) have also been cloned and shown to stabilize plasmids containing other *ARS* sequences. Therefore, when one desires stability and a single copy of a cloned gene, an *ARS* vector (Figure 2.1) containing a functional yeast centromere may be ideal.

In order to maximize expression of most foreign genes in yeast, transcription and translation should be controlled by yeast regulatory signals. Many expression vectors have been constructed to meet this requirement (see Hitzeman *et al.*, 1982; 1983; Reipen *et al.*, 1982; Tuite *et al.*, 1982) and examples of fusion vectors are described in Section V of this chapter. Figure 2.4 illustrates the expression vector YEp1PT (Hitzeman *et al.*, 1982) which has been used for the cloning and expression of the human interferon gene. This plasmid contains the yeast phosphoglycerate kinase (*PGK*) 5′ regulatory signals, a gene for selection in yeast (*TRP1*), a 2-μm DNA origin of replication and termination/polyadenylation signals, plus a pBR322 component for replication in bacteria. Genes for glycolytic enzymes are useful sources of portable promoter fragments, as they are easily regulated by the carbon source present in the growth medium.

METHOD 1:   CLONING THE *LEU2* GENE

Gene cloning procedures described below are illustrated by the cloning of the *LEU2* gene of this organism. The basic approach is as follows. Total yeast DNA is cut with the *Bam*HI restriction endonuclease and ligated to *Bam*HI-digested YRp7 DNA (see Figure 2.1). Since the *LEU2* gene of *S. cerevisiae* can complement the *leuB* mutation of *E. coli*, the ligated mixture of DNA is used to transform a *leuB* mutant of *E. coli* to a Leu⁺ phenotype. Transformants are then screened for plasmid DNA and potential recombinant plasmids are used to transform a double *leu2* mutant of *S. cerevisiae*.

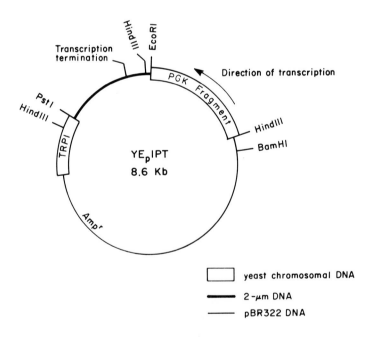

FIGURE 2.4   Partial map of the expression vector YEp1PT. Foreign DNA is cloned into the single *Eco*RI site, where it is under the control of the *PGK* promoter fragment. Abbreviations are the same as those in Figure 2.1. From Hitzeman *et al.*, 1982.

1. Mix 10–15 μg of *S. cerevisiae* chromosomal DNA with 2–4 μg of YRp7 DNA in *Bam*HI restriction buffer (see Chapter 1, Section IX, p. 36). Add 25 units of *Bam*HI restriction endonuclease and incubate at 37°C for 4 hours. Total volume = 50 μl.

2. Add 500 μL of 0.3M sodium acetate to the DNA.

3. Add an equal volume of TE buffer-saturated phenol, mix by inversion and spin for 10 minutes in a microfuge.

4. Carefully remove the supernatant (DNA) and combine with an equal volume of TE buffer-saturated ether. Mix and spin in a microfuge to separate the phases. Remove and discard the upper ether phase and repeat the extraction three or four times.

5. Remove residual ether by gentle evaporation under a stream of air.

6. Add 1.0 ml of chilled ethanol to the DNA, mix gently, and chill at −70°C for 5–15 mintues.

7. Pellet the DNA fragments by centrifugation for 10–15 minutes, pour off the supernatant and dry the pellet.

8. Add 20 μL of ligation buffer (see Chapter 1, Section X, p. 50) to the DNA pellet, dissolve, and add 2–3 units of T4 DNA ligase. Incubate overnight at 12–15°C.

9. Transform *E. coli* C600 using the procedure described in Chapter 1, Section XIV, p. 81 and spread cells on minimal medium supplemented with threonine and thiamine. Incubate the plates at 37°C for 2–3 days.

9. In our hands this procedure yields approximately 20 *LEU*$^+$ transformants.

Clones containing recombinant plasmids can be identified and characterized by a number of methods, most of which have been described elsewhere in this volume. In this specific case, recombinant colonies should be phenotypically Ap^r/Tc^s/Leu$^+$, and this phenotype should be transferable to other *E. coli* strains (by transformation). Physically, the recombinant plasmids should yield two fragments upon digestion with *Bam*HI, one corresponding to the vector YRp7 and the other to the *LEU2* DNA.

III. Transformation

A. HOST STRAINS

A number of yeast strains have been used as hosts in gene cloning experiments. The host strain chosen for a particular experiment will depend largely upon the cloning vector in use, since the selective markers on yeast vectors are genes that complement auxotrophic mutations present in the host, and in the gene to be cloned. In general, the following points should be taken into account.

(a) The strain should be readily transformable at a relatively high frequency. Variations in the transformation efficiency of different recipients have been observed (Johnston *et al.*, 1981). There are a large number of useful strains available from the Yeast Genetic Stock Centre (Donner Laboratory, University of California, Berkeley, CA 94720, U.S.A.; telephone (415) 843-2740, extension 6222 or (415) 642-2905, ask for Rebecca Contopoulou) and the American Type Culture Collection (12301 Parklawn Drive, Rockwell, MD 20852, U.S.A.).

(b) The mutation(s) present in the host strain, which can be complemented by gene(s) present on the cloning vector, should revert at a low frequency such that the identification of transformants is not complicated by the presence of prototrophic revertants. For 2-μm DNA or *ARS* vectors, point mutations are stable enough, but for an integrating vector it is useful to employ strains carrying either a deletion or two point mutations in the gene of interest.

(c) For biological containment, Botstein *et al.* (1979) constructed a series of sterile host strains (SHY), none of which are able to mate with other yeasts. These strains contain *ura3-52*, *trp1-289*, *leu2-3*, *leu2-112* and *his3-Δ1* mutations which can be complemented by genes on the appropriate vectors.

(d) Many strains of *S. cerevisiae* contain 2-μm DNA and although such strains can be transformed efficiently with 2-μm DNA-containing cloning vectors, the vector DNA is often unstable when cells are grown under nonselective conditions. One explanation for such instability is that endogenous 2-μm DNA may compete with incoming chimeric DNA during replication, or may recombine with the vector DNA. An alternative approach is to cure a *cir*$^+$ strain of endogenous 2-μm DNA by the procedure described by Erhart and Hollenberg (1981), and use this strain as a recipient for gene cloning.

METHOD 1: USE OF PROTOPLASTS

PROCEDURE

1. Grow a Leu$^-$ strain of *S. cerevisiae*, such as LL20, overnight, 30°C, in 30 ml of YEPD, to a density of $0.5-1 \times 10^8$ cells/mL.

2. Harvest the cells by low speed centrifugation and wash once with 10 mL distilled water.

3. Resuspend the cells in 15 mL of DTT buffer and incubate at 30°C, 10 minutes, with gentle shaking.

4. Wash twice with 10 mL of 1.2M sorbitol and resuspend the cells in 15 mL of the appropriate lytic enzyme buffer.

COMMENTS

1. Strain LL20 is *leu2-3 leu2-112 his3-11 his3-15* (constructed by G.R. Fink). The spontaneous mutation frequency to Leu$^+$ or His$^+$ is less than 10^{-10}.

3. DTT Buffer:

1.2M sorbitol
25 mM EDTA, pH 8.0
50 mM dithiothreitol (DTT).

The DTT should be added to the other components immediately prior to use and the solution filter sterilized.

4. Lytic Enzyme Buffer (for use with Glusulase):

1.2M sorbitol
5 mM dithiothreitol

| PROCEDURE | COMMENTS |

Prepare and filter sterilize just prior to use.

Lytic Enzyme Buffer (for use with Helicase):

1.2M sorbitol
25 mM EDTA, pH 8.0
50 mM dithiothreitol

5. Remove 0.1 mL of cells and determine cell number by spreading on minimal medium supplemented with histidine and leucine. To the remainder of the culture in lytic buffer, add Glusulase to a final concentration of 1% (v/v) or Helicase (0.1%), incubate 20 minutes, 30°C, with gentle shaking.

5. Minimal Medium:

0.67% yeast nitrogen base (no amino acids)
2% dextrose
20 μg/mL of amino acids
2% agar

6. Check protoplast formation by diluting 50 μL of the cell suspension in 1 mL 1.2M sorbitol and in 1 mL H$_2$O. The dilution in H$_2$O should clarify.

6. The efficiency of **PROTOPLAST FORMATION** can be monitored as described in the yeast DNA extraction procedure but in this case does not have to be complete.

7. Harvest the protoplasts by centrifugation at 2000 rpm, 5 minutes.

8. Wash three times in 1.2M sorbitol. Do *not* vortex the protoplasts.

8. The protoplasts must be washed thoroughly following Glusulase treatment, as some batches of enzyme contain DNase activity capable of digesting the transforming DNA.

9. Resuspend in 0.25 mL of 1.2M sorbitol, 10 mM CaCl$_2$, pH 8.0. Remove 10 μL in order to determine the number of intact cells (spread on minimal medium + histidine + leucine). Remove a second 10 μl sample, dilute to 10^{-4}, 10^{-5}, and add to 5 mL regeneration agar. Overlay on minimal medium + histidine + leucine (regeneration frequency).

9. Regeneration Agar (kept at 48°C):

1.2M sorbitol
0.67% yeast nitrogen base (no amino acids)
2% dextrose
2% agar
20 μg/mL amino acids (when necessary)

Regeneration agar is kept at 48°C to prevent solidification. Overlay on plates prewarmed at 37°C. Carry out all dilutions in 1.2M sorbitol.

10. Divide the remaining protoplasts into two 100 μL samples. To one, add approximately 5 μg of DNA isolated from the YRp7 *LEU2* recombinant plasmid. To the second, add 1.2M sorbitol.

10. Treat both samples the same and leave them for 15 minutes at room temperature. The transforming DNA should be in 1.2M sorbitol. Calculate the transformation frequency per regenerated protoplast.

11. Add 1.0 mL of 20% polyethylene glycol 4000, 10 mM Tris-HCl, 10 mM CaCl$_2$, pH 7.5 to the mixture and leave at room temperature for 20 minutes.

11. The protoplasts can be plated immediately after step 11 if desired. Proceed to step 14 but dilute 10× less.

12. Centrifuge at 2500 rpm, 5 minutes.

13. Add 100 μL of 1.2 M sorbitol, 10 mM CaCl$_2$, pH 8.0, + 50 μL of YEPD. Incubate 20 minutes, 30°C.

14. Prepare 10^{-1} and 10^{-2} dilutions in 1.2 M sorbitol, and add 100 μL samples of each to 5 mL regeneration agar. Overlay on minimal medium + histidine plates. Incubate 30°C, 3−5 days.

14. See comment 9.

METHOD 2:   USE OF ALKALI CATIONS

The procedure described below was recently developed (Ito *et al.*, 1983) and offers a rapid alternative to the procedure described above. It is applicable to *Schizosaccharomyces pombe* (R. Irving, N. Barton and A. Nasim, unpublished observations) as well as *S. cerevisiae*. Its application may be somewhat limited in that transformation efficiency is usually much lower than that found with protoplasts.

PROCEDURE	COMMENTS
1. Grow cells in YEPD medium to a density of $5-7 \times 10^6$/mL. Harvest by centrifugation for 5 minutes, 5000 rpm in the GSA rotor of a Sorval centrifuge.	**1.** It is best to determine the appropriate growth conditions which yield $5-7 \times 10^6$ cells/mL for each strain used. We do not advise using a spectrophotometric value for cell density unless each strain has been calibrated.
2. Wash once with an equal volume of sterile distilled water, collect cells by centrifugation and resuspend the pellet in 1/200 volume of 10 mM Tris-HCl, 1 mM EDTA, 0.1M lithium acetate, pH 7.5.	
3. Incubate at 30°C for 1.0 hour with agitation.	
4. Mix in a microfuge tube, 0.1 mL of cells with 40 μg of sonicated carrier DNA, and transforming DNA. Incubate 30 minutes at 30°C.	
5. Add 0.7 mL of 10 mM Tris-HCl, 1 mM EDTA, 0.1M lithium acetate, pH 7.5, 40% polyethylene glycol 4000, and incubate at 30°C for a further 30 minutes.	
6. Heat shock at 42°C for 5 minutes. Centrifuge for 20 seconds in a microfuge and wash the cells twice with 10 mM Tris-HCl, 1 mM EDTA, pH 7.5.	
7. Resuspend in 10 mM Tris-HCl, 1 mM EDTA, pH 7.5, and spread on selective plates.	**7.** This procedure does not work with all strains and can yield approximately 600 transformants per μg of DNA. Cell survival is approximately 40%.

IV. Selection of Clones

Since the *S. cerevisiae* recipient strains usually carry low reverting auxotrophic mutations, virtually all colonies which appear following transformation should carry the appropriate plasmid. This can be confirmed by a modification of the colony hybridization procedure described in Chapter 1, p. 65.

METHOD 1:   COLONY HYBRIDIZATION

PROCEDURE	COMMENTS
1. Place a sterile 0.45 μm nitrocellulose filter on a minimal agar plate supplemented with histidine (no	**1.** It is possible to replicate at least 100 colonies per filter.

leucine). Replicate potential transformants to the filter and to a master plate, using sterile toothpicks. Be sure the cells are rubbed firmly on the filter.

2. Mark the filter and master plate for orientation purposes. Incubate at 30°C, 12–16 hours.

2. Filters can be marked with pencil, or by applying unlabeled probe DNA in a specific pattern for each filter. The latter method has the advantage of measuring the efficiency of hybridization to the labeled probe DNA.

3. Remove the filter from the plate with forceps and place successively, colony side up, on several sheets of Whatman 3 MM paper saturated with the following solutions:

3. Yeast cells, unlike bacteria, must be treated with a reducing agent such as β-mercaptoethanol or dithiothreitol, and an enzyme mixture such as Glusulase, prior to lysis. This is due to the nature of the cell wall.

a. 50 mM EDTA, 2.5% β-mercaptoethanol, pH 9.0. Leave for 15 minutes.

a. Between the steps, touch the filter briefly on dry filter paper to remove excess liquid.

b. 1M sorbitol, 50 mM EDTA, pH 7.5, for 5 minutes.

c. 5 mg/mL Zymolyase 5,000 (Kirin Brewery) or 1% Glusulase in 1M sorbitol, 50 mM EDTA, pH 7.5. Incubate 2–3 hours at 37°C.

d. Transfer to 3 MM paper saturated with 0.2M NaOH, 0.6M NaCl, 1 minute.

d. The NaOH lyses the protoplasts and denatures the DNA. Single-stranded DNA is bound by nitrocellulose with a much greater efficiency than double-stranded DNA.

e. Transfer to 3 MM paper saturated with 1.0M Tris-HCl, pH 7.5, 1.5M NaCl, 5 minutes.

e. Washing with Tris/NaCl neutralizes the filter.

f. Wash in 2× SSC, for 5 minutes. Repeat once.

4. Air dry briefly and bake filter for 2–20 hours at 80°C.

4. Baking at 80°C fixes the DNA to the filter.

5. Hybridize with ^{32}P-labeled YRp7 *LEU2* DNA prepared by nick translation as described in Chapter 1, Section XII, p. 68 and autoradiograph (Chapter 1, Section XIII, p. 79).

5. ^{32}P-labeled probe DNA (YRp7 *LEU*2) is prepared as described in Chapter 1, Section XII, p. 67. See Chapter 1, Section XI, p. 59, for hybridization and Chapter 1, Section XIII, p. 79, for autoradiography procedures. The above procedure is essentially that of Hinnen *et al.*, (1978).

V. Gene Fusion Systems

In many cases, the study of the regulation of a particular gene is hampered by either a cumbersome assay for enzymatic activity of the gene product or the lack of a selectable phenotype associated with regulatory mutations. A number of systems have been employed with yeast to circumvent these problems by fusing the 5′ regulatory sequences of one gene to a protein-coding segment of a second gene encoding an easily assayable function and/or one for which mutants can be selected. One such system used in several studies to date (Guarente and Ptashne, 1981; Rose *et al.*, 1981), employs the β-galactosidase gene of *E. coli*. This gene encodes an enzyme which can be easily assayed in cell extracts and detected in yeast colonies. The expression of activity in yeast is dependent upon the placement of a yeast promoter and translational start signals upstream from the β-galactosidase coding sequences and thus makes β-galactosidase expression dependent upon the yeast gene regulatory sequences. Unfortunately, this system cannot be used for the selection of regulatory mutants due to the inability of *S. cerevisiae* to utilize lactose as an energy source.

A second system recently developed (Rymond *et al.*, 1983) depends upon the expression of the *E. coli* galactokinase gene in yeast. Galactokinase catalyzes the formation of galactose-1-phosphate from ATP and galactose as the first step in galactose catabolism in both *E. coli* and yeast cells, and yeast mutants deficient in galactokinase function, *gal1*, are unable to grow on minimal media with galactose as the sole energy source. This deficiency can be complemented by the transformation of such mutants with the *E. coli* galactokinase coding sequence fused to a yeast promoter, thus giving a selectable phenotype to the fusion gene. Since the expression of galactokinase activity is dependent upon the yeast regulatory sequences, the ability of such transformants to grow on galactose is also dependent upon these regulatory sequences. So, for example, when a plasmid containing the regulatory sequences of the yeast iso-1-cytochrome c gene, *CYC1*, fused to the galactokinase coding sequence was transformed into *gal1* cells, the resulting transformants were capable of growth on galactose in oxygen, but under anaerobic conditions which represses *CYC1* expression, these transformants could not grow. Mutants were selected which were capable of anaerobic growth on galactose and were constitutive for expression of both the fusion gene and the intact *CYC1* gene, thereby demonstrating the usefulness of this type of approach in yeast.

Fusions involving the galactokinase gene have been carried out using specially constructed galactokinase coding fragments (Figure 2.5) in two ways. First, protein coding fusions join the 5′ regulatory sequences plus the translational start codons of the yeast gene to the bulk of the galactokinase coding sequence. Such fusions can be achieved through an *Eco*R1 site placed in the beginning of the galactokinase coding sequence and require that the initiation ATG codon of the yeast gene be in-frame with the galactokinase coding sequence. This can be achieved either through the ligation of *Eco*R1 digested DNAs directly, or through the ligation of DNAs in which the single-stranded ends have been digested with S1 or filled-in with DNA polymerase in a combination which would generate an intact reading frame. The second type of fusion can be carried out between the intact galactokinase coding sequence

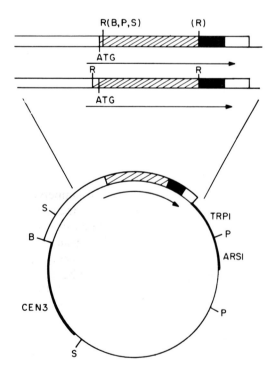

FIGURE 2.5 The *GalK* expression vectors. Two types of vectors are diagrammed; those which can be used for protein-coding fusions and those which can be used for mRNA leader sequence fusions. The filled-in boxes represent the *CYC1* coding sequences; the open boxes, the *CYC1* 5′ and 3′ noncoding sequences; the hatched boxes, the *GalK* sequences; the thick lines, other yeast sequences; and the thin line pBR322 sequences. The arrow designates the direction of transcription. Restriction enzyme sites are designated: B, *Bam*HI; P, *Pst*I; R, *Eco*RI; S, *Sal*I. Those in parentheses designate that vectors are available either with or wihtout those sites. The upper linear diagram represents a segment of those vectors suitable for protein coding fusions as the initiation ATG codon is 5′ to the useful restriction sites. The lower linear diagram represents a segment of those vectors suitable for mRNA fusions as the ATG initiation codon is 3′ to the useful restriction site.

preceeded by an *Eco*R1 site and the promoter and mRNA leader sequence of the yeast gene. This fusion employs the ATG codon of galactokinase and therefore places no requirements on the reading frame of the fusion. The type of fusion carried out depends on the availability of restriction sites within the protein coding sequence and the mRNA leader sequence. These fusions can be carried out most easily by the insertion of the yeast gene regulatory region into the galactokinase fusion plasmids outlined in Figure 2.5. These plasmids contain centromeres which are essential if a stable phenotype is to be maintained through mutant selections.

Fusion can be carried out using the following techniques.

METHOD 1: *IN VIVO* GALACTOKINASE ASSAY

PROCEDURE

1. Transform *gal1* cells as per Section III, p. 137.

2. Select transformants on plates containing 2% galactose, 0.67% yeast nitrogen base without amino acids (Difco) supplemented with appropriate growth factors.

COMMENTS

1. The isolation of clones containing galactokinase activity simply requires the transformation of *gal1* cells which are deficient in the endogenous yeast galacto-kinase activity.

2. Appropriate growth factors are determined by the nutritional requirements of the strain being transformed.

METHOD 2: *IN VITRO* GALACTOKINASE ASSAY

1. Prepare a crude cell extract by growing a 5 mL culture in selective medium to late exponential phase.

2. Chill the culture on ice and harvest by centrifugation (4°C).

3. Resuspend in cold distilled water and centrifuge again.

4. Resuspend the cells in 0.25 mL of lysis buffer, and transfer the suspension to a 1.5 mL microfuge tube. Add two-thirds volume of glass beads.

5. Mix the glass beads and cell suspension vigorously with a vortex mixer for 30 seconds and chill for 10 seconds.

6. Repeat this agitation/chilling cycle five additional times.

1. Selective medium is designed to maintain the presence of the plasmids.

4. **Lysis Buffer:**

   20 mM HEPES, pH 7.5 (with KOH)
   1 mM DTT
   200 μg/mL bovine serum albumin

   This buffer can be stored for two months.

6. Cell lysis may be monitored microscopically.

7. With a syringe needle, punch a hole in the bottom of the microfuge tube. Place this tube into the mouth of a glass centrifuge tube and spin in a table top centrifuge for a few seconds.

7. The extract passes quickly through the hole into the glass tube leaving the glass beads behind.

8. Wash the glass beads once with 0.25 mL of lysis buffer and spin again into the glass tube from step 7.

9. Clarify the extract by centrifugation for 2 minutes in a microfuge. Store the supernatant in 0.1 mL aliquots at $-60°C$, to $-80°C$.

10. To assay for enzyme activity, assemble an 80 μL reaction mix on ice for each amount of enzyme extract to be tested plus a blank control.

10. Reaction Mix (per 80 μL)

5 mM DTT

16 mM NaF (stored in single aliquots at $-60°C$ to $-80°C$)

10 μL 1.0M Tris-HCl, pH 7.9

20 μL 20 mM $MgCl_2$

20 μL 8 mM ATP

10 μL ^{14}C-galactose

Initially a stock solution of 100 μL of ^{14}C-galactose (50 μCi/μmole, 200 μCi/mL from Amersham), 10 μL of 1M galactose and 1 mL H_2O can be tried. The specific activity of ^{14}C-galactose can be varied depending upon the sensitivity required. More radioactive or unlabeled galactose can be added as desired.

11. Preincubate the 80 μL reaction mixture for 30 seconds at 32°C.

12. Add $0-20$ μL of cell extract (from step 9) plus 20 μL of lysis buffer to the reaction mix.

12. Assure a total volume of 100 μL for each extract concentration and control.

13. For each time point, remove 25 μL and add it immediately to 2 μL of stop solution in a tube, and store the tube on ice.

13. STOP Solution:

0.5M galactose

250 mM EDTA

Time points from $1-10$ minutes are recommended initially, but longer times can be used for increased sensitivity if required.

14. From each tube, remove 20 μL and apply it onto a DE81 (Whatman) filter disc.

15. After the spots are dried, place the filters into a beaker of ice cold 1% galactose.

15. Use a large beaker containing sufficient 1% galactose to cover the filter discs.

16. Change the wash solution four times, and follow with a fifth wash of H_2O.

17. Dry the filters and determine the radioactivity in a liquid scintillation counter.

17. For the calculations of pmoles of galactose-1-phosphate formed per minute per mg protein, use the initial slope of the curve (pmoles enzyme versus time).

VI. References

Beggs, J.D. 1978. Transformation of yeast by a replicating hybrid plasmid. *Nature* 275:104–109.

Botstein, D., S.C. Falco, S.E. Stewart, M. Brennan, S. Scherer, D.T. Stinchcomb, K. Struhl, and R.W. Davis. 1979. Sterile host yeasts (SHY): a eukaryotic system of biological containment for recombinant DNA experiments. *Gene* 8:17–24.

Broach, J.R. 1982. The yeast plasmid 2µ circle. *Cell* 28:203–204.

Broach, J.R., J.N. Strathern, and J.B. Hicks. 1979. Transformation in yeasts: development of a hybrid cloning vector and isolation of the *CAN1* gene. *Gene* 8:121–133.

Clarke, L., and J. Carbon. 1980. Isolation of a yeast centromere and construction of functional small circular chromosomes. *Nature* 287:504–509.

Crosby, W.L., and D.Y. Thomas. 1983. Gene cloning in yeast, 1–19. *In* R.A. Flavell (ed.), Techniques in Life Sciences, B5, Nucleic Acid Biochemistry, B503, Elsevier, New York.

Cryer, D.F., R. Eccleshall, and J. Marmur. 1975. Isolation of yeast DNA, 39–44. *In* D.M. Prescott (ed.), Methods in Cell Biology, Vol. XII, Academic, New York.

Devenish, R.J., and C.S. Newlon. 1982. Isolation and characterization of yeast ring chromosome III by a method applicable to other circular DNA's. *Gene* 18:277–288.

Erhart, E., and C.P. Hollenberg. 1981. Curing of *Saccharomyces cerevisiae* 2-µm DNA by transformation. *Curr. Genet.* 3:83–89.

Guarente, L., and M. Ptashne. 1981. Fusion of *Escherichia coli lac2* to the cytochrome C gene of *Saccharomyces cerevisiae*. *Proc. Natl. Acad. Sci. U.S.A.* 78:2190–2195.

Guerineau, M. 1979. Plasmid DNA in yeast, 539–593. *In* P.A. Lemke (ed.), Viruses and Plasmids in Fungi, Series on Mycology, Vol. 1, Marcel Dekker, New York.

Hawthorne, D.C., and R.K. Mortimer. 1978. Genetic map of *Saccharomyces cerevisiae*. Handbook of Biochemistry and Molecular Biology 2:765–832.

Hicks, J.B., A. Hinnen, and G.R. Fink. 1979. Properties of yeast transformation. Cold Spring Harb. Symp. Quant. Biol. 43:1305–1313.

Hinnen, A., J.B. Hicks, and G.R. Fink. 1978. Transformation of yeast. *Proc. Natl. Acad. Sci. U.S.A.* 75:1929–1933.

Hinnen, A., and Meyhack, B. 1982. Vectors for cloning in yeast. *Curr. Top. Microbiol. Immunol.* 96:101–117.

Hitzeman, R.A., D.W. Leung, L.J. Perry, W.J. Kohr, H.L. Levine, and D.V. Goeddel. 1983. Secretion of human interferons by yeast. *Science* 219:620–625.

Hitzeman, R.A., D.W. Leung, L.J. Perry, W.J. Kohr, F.E. Hagie, C.Y. Chen, J.M. Lugovoy, A. Singh, H.L. Levine, R. Wetzel, and D.V. Goeddel. 1982. Expression, processing, and secretion of heterologous gene products by yeast. *Rec. Adv. Yeast Mol. Biol.* 1:173–190.

Hollenberg, C.P. 1982. Cloning with 2-µm DNA vectors and the expression of foreign genes in *Saccharomyces cerevisiae*. *Curr. Top. Microbiol. Immunol.* 96:109–144.

Hsiao, C.L., and J. Carbon. 1981. Direct selection procedure for the isolation of functional centromeric DNA. *Proc. Natl. Acad. Sci. U.S.A.* 78:3760–3764.

Hyman, B.C., J.H. Cramer, and R.H. Rownd. 1982. Properties of a *Saccharomyces cerevisiae* mt DNA segment conferring high frequency yeast transformation. *Proc. Natl. Acad. Sci. U.S.A.* 79:1578–1582.

Ito, H,. Y. Fukuda, K. Murata, and A. Kimura. 1983. Transformation of intact yeast cells treated with alkali cations. *J. Bacteriol.* 153:163–168.

Johnston, J., F. Hilger, and R. Mortimer. 1981. Variation in transformation frequency by plasmid YRp7 in *Saccharomyces cerevisiae*. *Gene* 16:325–329.

Kingsman, A.J., L. Clarke, R.K. Mortimer, and J. Carbon. 1979. Replication in *Saccharomyces cerevisiae* of plasmid pBR313 carrying DNA form the yeast *TRP1* region. *Gene* 7:141–153.

Livingston, D.M. 1977. Inheritance of the 2 μm DNA plasmid from *Saccharomyces. Genetics* 86:73−84.

Lörincz, A.T., and S.I. Reed. 1984. Primary structure homology between the product of yeast cell division control gene *CDC28* and vertebrate oncogenes. *Nature* 307:183−185.

McAleer, W.J., E.B. Buynak, R.Z. Maigetter, D.E. Wampler, W.J. Miller, and M.R. Hilleman. 1984. Human hepatitis B vaccine from recombinant yeast. *Nature* 307:178−180.

Murray, A.W., and J.W. Szostak. 1983. Construction of artificial chromosomes in yeast. *Nature* 305:189−193.

Orr-Weaver, T.L., J.W. Szostak, and R.J. Rothstein. 1981. Yeast transformation: A model system for the study of recombination. *Proc. Natl. Acad. Sci. U.S.A.* 78:6354−6358.

Reipen, G., E. Erhart, K.D. Breunig, and C.P. Hollenberg. 1982. Nonselective transformation of *Saccharomyces cerevisiae. Curr. Genet.* 6:189−193.

Roeder, G.S., and G.R. Fink. 1980. DNA rearrangements associated with a transposable element in yeast. *Cell* 21:239−249.

Rose, M., M.J. Casadaban, and D. Botstein. 1981. Yeast genes fused to β-galactosidase in *Escherichia coli* can be expressed normally in yeast. *Proc. Natl. Acad. Sci.* U.S.A. 78:2460−2464.

Rymond, B.C., R.S. Zitomer, D. Schumperl, and M. Rosenberg. 1983. The expression in yeast of the *Escherichia coli GalK* gene on *CYCI/GalK* fusion plasmids. *Gene* 25:249−262.

Scherer, S., and R.W. Davis. 1979. Replacement of chromosomal segments with altered DNA sequences constructed in vitro. *Proc. Natl. Acad. Sci. U.S.A.* 76:4951−4955.

Seligy, V.L., D.Y. Thomas, and B.L.A. Miki. 1980. *Saccharomyces cerevisiae* plasmid, Scp or 2 μm: intracellular distribution, stability and nucleosomal-like packaging. *Nucl. Acids Res.* 8:3371−3392.

Singh, H., J.J. Bieker, and L.B. Dumas. 1982. Genetic transformation of *Saccharomyces cerevisiae* with single-stranded DNA vectors. *Gene* 20:441−449.

Skipper, N., D.Y. Thomas, and P.C.K. Lau. 1984. Cloning and sequencing of the pre protoxin-coding region of the yeast M1 double-stranded RNA. *EMBO Journal* 3:107−111.

Stinchcomb, D.T., C. Mann, and R.W. Davis. 1982. Centromeric DNA from *Saccharomyces cerevisiae. J. Mol. Biol.* 158:157−179.

Stinchcomb, D.T., K. Struhl, and R.W. Davis. 1979. Isolation and characterization of a yeast chromosomal replicator. *Nature* 282:39−43.

Storms, R.K., J.B. McNeil, P.S. Khandakar, G. An, J. Parker, and J.D. Friesen. 1979. Chimeric plasmids for cloning of deoxyribonucleic acid sequences in *Saccharomyces cerevisiae. J. Bacteriol.* 140:73−82.

Strathern, J.N., E.W. Jones, and J.R. Broach (ed.), 1981. The molecular biology of the yeast *Saccharomyces*: life cycle and inheritance. Cold Spring Harbor Laboratory, New York, 751.

Strathern, J.N., E.W. Jones, and J.R. Broach, (ed.) 1982. The molecular biology of the yeast *Saccharomyces*: metabolism and gene expression. Cold Spring Harbor Laboratory, New York, 680.

Struhl, K. 1983. The new yeast genetics. *Nature* 305:391−397.

Struhl, K., D.T. Stinchcomb, S. Scherer, and R.W. Davis. 1979. High frequency transformation of yeast: autonomous replication of hybrid molecules. *Proc. Natl. Acad. Sci. U.S.A.* 76:1035−1039.

Tuite, M.F., M.J. Dobson, N.A. Roberts, R.M. King, D.C. Burke, S.M. Kingsman, and A.J. Kingsman. 1982. Regulated high efficiency expression of human interferon-alpha in *Saccharomyces cerevisiae. EMBO J.* 1:603−608.

Wu, R., L. Grossman, and K. Moldave (ed.), 1983. Methods in Enzymology, vol. 101, Part C—Recombinant DNA. Academic, New York.

chapter 3

GENE CLONING IN *PSEUDOMONAS AERUGINOSA*

A.A. Potter

I. Introduction

Pseudomonas aeruginosa is a gram-negative nonspore-forming aerobe, characterized in many cases by production of a water-soluble blue-green pigment. This organism is important from both medical and environmental viewpoints. Medically, it has been a major source of nosocomial infection and is a major cause of pulmonary complications in cystic fibrosis patients. Environmentally, *P. aeruginosa*, like many other Pseudomonads, is able to degrade a wide range of organic compounds for use as sources of carbon and nitrogen. *P. aeruginosa* is well characterized genetically (Clarke and Richmond, 1975; Holloway *et al.*, 1979) and procedures have been developed for genome analysis by conjugation, transduction, and transformation. Since many *P. aeruginosa* genes are not fully expressed in *Escherichia coli*, the development of gene cloning systems for use in *P. aeruginosa* is an ongoing concern in many laboratories. The observations that large overproduction of *E. coli* proteins is possible by introducing the cloned gene into *P. aeruginosa* (Sakaguchi, 1982) makes gene cloning in this organism commercially significant. For more traditional methods of genetic analysis in this organism, the reader is referred elsewhere (Clarke and Richmond, 1975, Holloway *et al.*, 1979).

A. CLONING VECTORS

Common plasmid vectors used in *E. coli*, such as pBR322 and pACYC184, are not stably maintained in *P. aeruginosa*. There are some indigenous plasmids which may be useful for the development of cloning vehicles, but to a large degree they are not sufficiently characterized to be useful at present. One exception, plasmid RP1, is discussed below. The plasmid RK2 (very similar if not identical to RP1) has served as the source of a number of very useful cloning vectors which have been used extensively with a variety of gram-negative hosts. This cloning system has been described in detail elsewhere (Ditta *et al.*, 1980) and will not be included here. Most of the cloning vectors described in this chapter are based on the RSF1010 replicon. RSF1010 is a small (8.9 Kb) nonconjugative member of the P-4 incompatibility group, specifying resistance to sulphonamide (Su) and streptomycin (Sm) (Barth and Grinter, 1974). This plasmid is stably maintained in both *P. aeruginosa* and *E. coli* and can be mobilized for transfer by coexisting conjugative plasmids. RSF1010 itself is a poor cloning vector due to its lack of unique restriction endonuclease cleavage sites and lack of suitable selective markers. Sulphonamide resistance is not a convenient marker to use due to high levels of resistance exhibited by some *P. aeruginosa* strains (Potter, 1981). However, a number of RSF1010 derivatives have been constructed which contain additional restriction endonuclease cleavage sites and resistance genes (Bagdasarian and Timmis, 1982; Bagdasarian *et al.*, 1981; Sakaguchi, 1982). Four such derivatives are illustrated in Figure 3.1.

Vectors pKT210 and pKT248 carry chloramphenicol-resistance genes from plasmids S-a and R621a1a, respectively, and can be used to clone *Hin*dIII, *Sal*I, *Eco*RI and *Sst*I-generated DNA fragments. Plasmid pKT231 contains the neomycin-resistance gene from transposon Tn*903* and is useful for cloning *Eco*RI, *Sst*I, *Hin*dIII, *Xho*I and *Sma*I-generated DNA fragments. The plasmid pKT247 contains the λ *cos* site and ampicillin-resistance gene of Tn*3*, and contains unique sites for *Eco*RI and *Sst*I. This plasmid is useful for the construction of *P. aeruginosa* gene banks, using the λ *in vitro* packaging system in *E. coli* (Collins and Hohn, 1978) and subsequent transfer to *Pseudomonas* by mobilization or transformation. The reader is referred

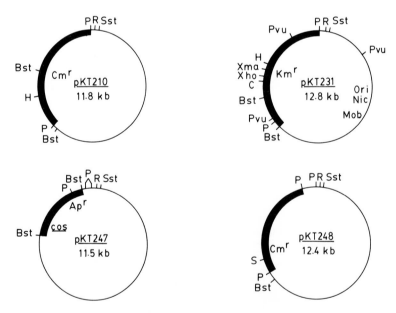

FIGURE 3.1 Restriction endonuclease maps of RSF1010-based cloning vectors pKT210, pKT231, pKT247, and pKT248. The thin portion of the circle represents the RSF1010 replicon while the thick portion represents selectable markers from other sources (see text for description). Abbreviations: Sm[r], streptomycin resistance; Cm[r], chloramphenicol resistance; Km[r], kanamycin resistance; Ap[r], ampicillin resistance, *cos*, *cos* region from bacteriophage lambda, Ori, origin of replication; Nic, the relaxation nick site; Mob, ability to be mobilized for transfer by conjugative plasmids. Restriction endonuclease cleavage sites are as follows: P, *Pst*I; R, *Eco*RI; Sst, *Sst*I; Bst, *Bst*EII; H, *Hin*dIII; Pvu, *Pvu*II; C, *Cla*I; Xho, *Xho*I; Xma, *Xma*I. From Bagdasarian and Timmis (1982).

GENE CLONING IN *PSEUDOMONAS AERUGINOSA*

to Bagdasarian and Timmis (1982), and Bagdasarian *et al.*, (1981) for further details on these and other cloning vectors.

Olsen *et al.*, (1982) constructed cloning vectors based on the wide host range plasmid RP1. A deletion derivative of RP1, pR01600, was isolated following conjugation and a Tn*3* insertion mutant of the latter plasmid, pR01601, was subsequently isolated. Finally, pR01601 was digested with *Pst*I to eliminate nonessential regions, and then transformed into *P. aeruginosa* following self-ligation or ligation with pBR322. The resulting plasmids, pR01613 and pR01614, are illustrated in Figure 3.2 and have been shown to be useful in cloning *Pst*I and *Bam*HI or *Hin*dIII fragments of the *P. aeruginosa* chromosome. These plasmids are not ideal vectors in that pR01613 loses its ampicillin-resistance marker when cleaved with *Pst*I and hence can only be used when direct selection for the cloned DNA is possible. Plasmid pR01614, containing pBR322, utilizes cloning sites within the tetracycline resistance (Tc^r) gene. High copy number Tc^r plasmids have been shown to be unstable in the absence of selection in *P. aeruginosa* (Wood *et al.*, 1981; Bagdasarian and Timmis, 1982) and therefore the use of insertional inactivation of Tc^r may be unsuitable for most purposes. Despite this, both plasmids have proven useful in the construction of *P. aeruginosa* gene banks (Olsen *et al.*, 1982).

B. *P. AERUGINOSA* HOST STRAINS

Host strains used for gene cloning should be readily transformable, lack a functional restriction system, and exhibit reduced general recombination. Many *P. aeruginosa* strains are readily transformable, and restrictionless *(rmo)* and *recA$^-$* mutants are available (Dunn and Holloway, 1971; Chandler and Krishnapillai, 1974; see Table 3.1). Strains carrying both *recA* and the *rmo* mutations have not been constructed. However, the *recA* mutation, which determines homologous recombination, is generally not needed when cloning foreign DNA fragments and therefore one can use the restrictionless strain PA01162. Similarly, the *rmo* genotype is not necessary for the cloning of *P. aeruginosa* DNA and the *recA* mutant PA02003 can be used successfully.

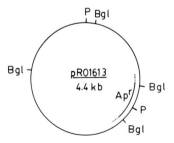

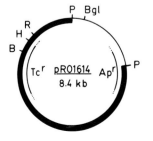

FIGURE 3.2 RP1-based cloning vectors pR01613 and pR01614. The thin line represents RP1 DNA, while the thick line represents DNA of pBR322. See text for description. Abbreviations: same as Figure 3.1 and Bgl, *Bgl*I. Based on Olsen *et al.*, (1982).

P. aeruginosa strain 1 (PAO) contains an exonuclease capable of degrading double-stranded linear DNA (Potter and Loutit, 1982). Although this enzyme does not play a role in normal plasmid cloning and transformation (Potter and Loutit, unpublished observations), it is able to degrade linear chromosomal or phage DNA during transformation. Mutants defective in this nuclease, termed *res⁻* since they were originally thought to be restrictionless, have been described (Mercer and Loutit, 1978, 1979; Potter and Loutit, 1982) and are transformable with chromosomal DNA. Therefore, it is possible to subject a region of cloned chromosomal DNA to *in vitro* mutagenesis and introduce the mutagenized segment into *P. aeruginosa* by transformation of a *res⁻* strain. Transformants can then be analyzed to assess the effect of the mutation. Since the enzyme specified by the *res⁺* gene does not degrade DNA with single-stranded terminal regions, as was shown for phage F116 (Potter and Loutit, 1982), it should be possible to carry out the same procedures using a wild-type strain if the mutagenized region is cut out with a restriction endonuclease which produces "sticky" ends. This is illustrated in Figure 3.3. In either case, the recipient strain would have to be *recA⁺*, as transformants are detected only after integration into the host chromosome.

Although the methods for gene cloning and for the detection of recombinants in Pseudomonads are the same as described in Chapter 1, plasmid isolation and transformation procedures are slightly modified in *Pseudomonas* species. They are described below.

TABLE 3.1 *P. aeruginosa* Strains Useful for Gene Cloning

Strain	Relevant Characteristics	Reference
PA01162	*leu-38 rmo-11* (restrictionless)	Dunn and Holloway, 1971
PA02003	*argH⁻ recA⁻*	Chandler and Krishnapillai, 1974
OT684	*leu-1 lys-1 res-4*	Potter and Loutit, 1982

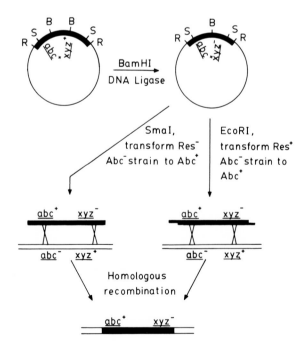

FIGURE 3.3 Procedure for the replacement of a chromosomal gene with a DNA sequence subjected to *in vitro* mutagenesis. A recombinant plasmid carrying two hypothetical genes, *abc⁺* and *xyz⁺*, is mutagenized within *xyz⁺* by deletion of a *Bam*HI fragment. The cloned chromosomal fragment carrying *abc⁺ xyz⁻* is then excised with *Eco*RI ("sticky" ends) or *Sma*I (blunt ends), the fragments purified, and used to transform a *P. aeruginosa abc⁻ xyz⁺* strain to *abc⁺*. In the case of the *Eco*R1-generated fragment, any *recA⁺* recipient can be employed since the Res exonuclease will not degrade single-stranded DNA. For blunt-ended fragments, such as those generated by *Sma*I digestion, a *res-4* mutant must be used as a recipient chromosome at the *abc⁻ xyz⁺* site. This is based on work carried out using DNA of the bacteriophage F116 (Potter and Loutit, 1982) and chromosomal DNA fragments (Mercer and Loutit, 1978; Potter and Loutit, unpublished observations).

II. Purification of Plasmid DNA from *Pseudomonas* Isolates

Most plasmids can be easily purified from *P. aeruginosa* by modifications of methods developed for use with *E. coli*. The procedure described below is useful for the isolation of small to medium size plasmids (Mercer, 1979). Large plasmids such as FP2 (Loutit, 1970) can be extracted quantitatively by the alkaline denaturation procedure of Birnboim and Doly (1979).

METHOD 1: ISOLATION OF MEDIUM SIZE PLASMIDS BY CsCl-EtBr ULTRACENTRIFUGATION

PROCEDURE

1. Grow cells overnight in 500 mL of L broth, at 37°C in a gyratory water bath.

2. Harvest the cells by centrifugation at 10,000 rpm at 4°C for 10 minutes. Resuspend the pellet in 50 mL of ice-cold 10 mM Tris-HCl, pH 8.0.

3. Add 15 mL of chilled 0.25M EDTA, pH 8.0, and leave on ice for 5 minutes.

4. Add 10 mg of solid lysozyme. Mix well, and leave on ice for a further 5 minutes.

5. Add 10 mL of lysis buffer. Immediately centrifuge the mixture at 12,000 × g for 30 minutes at 4°C.

6. Collect the supernatant (cleared lysate). Add NaCl to a final concentration of 1M and polyethylene glycol 6000 to 10% (w/v). Incubate overnight at 0−4°C.

7. Harvest the precipitated DNA by centrifugation at 5,000 rpm for 15 minutes at 4°C and resuspend in 5−10 mL of 20 mM Tris-HCl, pH 8.0, 2 mM EDTA.

COMMENTS

1. For L broth recipe, see Chapter 1, Section XIV, p. 82. *P. aeruginosa* can be grown in standard micro-biological media such as brain heart infusion or nutrient broth (Difco Laboratories Inc.). The addition of potassium nitrate (8 g/L) prevents pellicle formation and facilitates growth ($2-4 \times 10^9$ cells/mL after overnight incubation at 37°C).

When preparing solid media from liquid media, the addition of agar (Difco) at a concentration of 2% (w/v) minimizes the problem of colony-spreading which is frequently encountered with *P. aeruginosa*.

The minimal medium of Davis and Mingioli (1950) may also be used; although citrate (0.2% w/v) should be substituted for glucose as a carbon source.

Amino acids are added to a final concentration of 0.1 mM with the exception of isoleucine and valine, which are added at 1.0 mM.

5. Lysis Buffer:

10 mM Tris-HCl, pH 8.0
62.5 mM EDTA
2% Triton X-100

8. Mix an equal volume of cold chloroform:octanol (98:2) and separate the phases by centrifugation at 8,000 × g, 15 minutes.

9. Remove the upper aqueous phase and add solid CsCl and ethidium bromide to give a density of 1.55 g/mL and a dye concentration of 500 µg/mL. Generally, this can be achieved by using the following ratios: 6.37 g CsCl + 6.45 g lysate + 0.425 mL ethidium bromide.

9. The chloroform octanol phase will be on the bottom while the aqueous phase will be on top. Ethidium bromide is stored as a 10 mg/ml solution in the dark.

10. Spin in the Type 50 Ti rotor of a Beckman ultracentrifuge at 45,000 rpm, 48 hours.

11. View the tube under ultraviolet illumination, and remove the lower plasmid band. Extract with CsCl-saturated isopropanol as described in Chapter 1, Section IV to remove the ethidium bromide.

12. Add 2 volumes of water and 0.1 volume of 3M sodium acetate. Precipitate the DNA by adding 2 volumes of ethanol and keep at −20°C for 8−24 hr.

13. Collect the DNA by centrifugation. Decant the ethanol and air dry the tube. Resuspend the DNA pellet in 0.5−1.0 mL of 20 mM Tris-HCl, pH 8.0, 2 mM EDTA.

METHOD 2: RAPID SCREENING PROCEDURES FOR PLASMID DNA

The presence of plasmid DNA in *P. aeruginosa* transformants can be screened for by a variety of procedures. The first method described below is useful for a number of *Pseudomonas* species (Kado and Liu, 1981) and yields plasmid DNA which can be digested with restriction endonucleases. The second method, which is very similar to that of Birnboim and Doly (1979), is useful for the isolation of large plasmids such as FP2 (Potter and Loutit, unpublished observations). This plasmid is 60 Md in size and specifies resistance to mercuric chloride. The method was developed by Portnoy, White, and Falkow, and was communicated to the author by John Loutit.

A. GENERAL METHOD

PROCEDURE   **COMMENTS**

1. Prepare an overnight culture in 0.2−1.0 mL of broth.

2. Harvest the cells in a 1.5 mL microfuge tube, and resuspend in 10 mM Tris-HCl, pH 8.0 (100 µL).

PROCEDURE	COMMENTS

3. Add 400 µL of lysing solution.

3. Lysing Solution

3% SDS in 50 mM Tris, pH 12.6

Adjust pH to 12.6 by adding 1.6 mL of 2M NaOH to 98.4 mL of Tris-SDS.

4. Incubate at 55°–60°C for 1.0 hour.

5. Add 2 volumes of phenol:chloroform (1:1), mix, and centrifuge to separate the phases. Remove the upper phase.

6. Mix with an equal volume of Tris-saturated ether (10 mM Tris-HCl, pH 8.0) and centrifuge briefly to separate the phases. Remove the upper phase (ether) and discard. Repeat 2 or 3 times to remove all traces of phenol.

7. Add 50 µL of 3M sodium acetate, followed by 800 µL of chilled ethanol (−20°C). Place in a dry ice/ethanol bath for 5–15 minutes.

7. If dry ice is unavailable, a −70°C freezer can be used effectively.

8. Centrifuge 15 minutes at 4°C, drain ethanol and dry the pellet. Resuspend in 50 µL of sterile distilled water.

9. Use 20 µL portions for analysis by agarose gel electrophoresis.

B. ISOLATION OF LARGE PLASMIDS

PROCEDURE	COMMENTS

1. Grow cells in 2 mL of Difco brain heart infusion broth to an optical density of 1.0–1.5 (560 nm). Harvest the cells by centrifugation (5 min, low speed).

2. Wash with 2 mL of TE. Harvest by centrifugation and resuspend in 50 µL of TE.

2. TE

50 mM Tris-HCl, pH 8.0
10 mM EDTA

3. Transfer the cells to a 1.5 mL microfuge tube containing 0.6 mL of lysis buffer (TE + 4% SDS, pH 12.42) and mix by rapid but gentle inversion of the tube.

4. Incubate at 37°C for up to 20 minutes to ensure complete lysis.

5. Neutralize to pH 8.0 by the addition of 50 µL of 2M Tris-HCl, pH 7.0. The viscosity should decrease rapidly as the solutions are mixed.

6. Precipitate single-stranded DNA by the addition of 0.16 mL of 5M sodium chloride. Chill on ice for 1 hour.

7. Centrifuge 5 minutes in a microfuge and decant the supernatant into a clean tube.

8. Precipitate the plasmid DNA by the addition of 0.55 mL of isopropanol. Store at $-70°C$ for $5-15$ minutes and collect the precipitate by centrifugation for 10 minutes in a microfuge.

9. Resuspend the pellet in the appropriate buffer.

9. If a restriction analysis is desired, resuspend the dried precipitate in 100 μL of TE buffer and extract once with phenol as described in the previous procedure (step 5). Precipitate the DNA by adding 10 μL of 3M sodium acetate and $2-3$ volumes of ethanol. Cool for 5 minutes at $-70°C$, collect the DNA by centrifugation and resuspend in 30 μL of TE buffer.

If no restriction enzyme digestion is desired, resuspend the pellet (step 8) in 30 μL of TE and analyze 10 μL portions by agarose gel electrophoresis.

III. Transformation of Pseudomonads

The transformation procedure described below is a modification of that described by Mercer and Loutit (1978, 1979) and can be used effectively for linear or circular DNA (Potter and Loutit, 1981). Bagdasarian and Timmis (1982) have recently described an alternate procedure using RbCl-CaCl$_2$ which yields more than 10^5 transformants per μg of DNA, comparable to this method.

1. Dilute 0.75 mL of an overnight broth culture (Difco brain heart infusion supplemented with 8 g/L potassium nitrate) into 10 mL of the same and incubate with gentle shaking for 4 hours at 37°C.

1. We routinely use a 30 mL bottle for cell growth in a gyratory water bath.

2. Harvest the cells by centrifugation at 10,000 rpm for 10 minutes at 4°C and resuspend in 5.0 mL of chilled 0.1 M MgCl$_2$.

2. As mentioned previously (Chapter 1, Section XIV, p. 81), prechill all tubes and pipettes.

3. Centrifuge at 10,000 rpm for 10 minutes and resuspend the cells in 5.0 mL of chilled 0.15M MgCl$_2$. Incubate for 20 minutes in an ice bath.

4. Centrifuge again. Resuspend the cells in 1 mL 0.15M MgCl$_2$. Dispense into 0.2 mL aliquots (use chilled tubes) for transformation.

5. Add up to 100 μL (3 μg) of DNA to the cells and incubate for $30-60$ minutes in an ice bath.

5. If chromosomal DNA is used, >3 μg can be added.

6. Heat pulse for 1 minute at 53°C or 2 minutes at 42°C. Leave at room temperature for 5 minutes.

7. Dilute by adding 3.0 mL of Difco brain heart infusion broth and incubate at 37°C for 90 minutes.

8. Plate on selective media.

8. Antibiotics and other substances should be used singly or in combination at the following concentrations in a rich solid medium (e.g., brain heart infusion or nutrient agar):

$$500 \; \mu g/mL \; \text{carbenicillin}$$
$$200 \; \mu g/mL \; \text{tetracycline HCl}$$
$$250 \; \mu g/mL \; \text{kanamycin}$$
$$500 \; \mu g/mL \; \text{streptomycin}$$
$$1000 \; \mu g/mL \; \text{spectinomycin}$$
$$250 \; \mu g/mL \; \text{rifampicin}$$
$$12 \; \mu g/mL \; \text{HgCl}_2$$

When streptomycin and $HgCl_2$ are used together, reduce the concentration of the latter 5-fold.

IV. References

Bagdasarian, M., and K.N. Timmis. 1982. Host:vector systems for gene cloning in *Pseudomonas*, 47–67. In P.H. Hofschneider, and W. Goebel (ed.), Current Topics in Microbiology and Immunology. Springer Verlag, Berlin.

Bagdasarian, M., R. Lurz, B. Ruckert, F.C.H. Franklin, M.M. Bagdasarian, J. Frey, and K.N. Timmis. 1981. Specific purpose plasmid cloning vectors II. Broad host range, high copy number, RSF1010—derived vectors, and a host-vector system for gene cloning in *Pseudomonas*. Gene 16:237–247.

Barth, P.T., and N.G. Grinter. 1974. Comparison of the deoxyribonucleic acid molecular weights and homologies of plasmids carrying linked resistance to streptomycin and sulfonamides. J. Bacteriol. 120:618–630.

Birnboim, H., and J. Doly. 1979. A rapid extraction procedure for screening recombinant plasmid DNA. Nucl. Acids Res. 7:1513–1525.

Chandler, P.M., and V. Krishnapillai. 1974. Isolation and properties of recombination-deficient mutants of *Pseudonomas aeruginosa*. Mutat. Res. 23:15–23.

Clarke, P.H., and M.H. Richmond. 1975. Genetics and biochemistry of *Pseudomonas*. John Wiley and Sons, London.

Collins, J., and B. Hohn. 1978. Cosmid: a type of plasmid gene-cloning vector that is packageable *in vitro* in bacteriophage heads. Proc. Natl. Acad. Sci. (USA) 75:4242–4246.

Davis, B.D., and E.S. Mingioli. 1950. Mutants of *Escherichia coli* requiring methionine or vitamin B12. J. Bacteriol. 60:17–28.

Ditta, G., S. Stanfield, D. Corbin, and D.R. Helinski. 1980. Broad host range DNA cloning system for gram-negative bacteria; construction of a gene bank of *Rhizobium meliloti*. Proc. Natl. Acad. Sci. 77:7347–7351.

Dunn, N.W., and B.W. Holloway. 1971. Pleiotropy of p-fluorophenylalanine-resistant and antibiotic hypersensitive mutants of *Pseudomonas aeruginosa*. Genet. Res. 18:185–197.

Holloway, B.W., V. Krishnapillai, and A.F. Morgan. 1979. Chromosomal genetics of *Pseudomonas*. Microbiol. Rev. 43:73–102.

Kado, C.I., and S.-T. Liu. 1981. Rapid procedure for detection and isolation of large and small plasmids. J. Bacteriol. 145:1365–1373.

Loutit, J.S. 1970. Investigation of the mating system of *Pseudomonas aeruginosa* strain 1. VI. Mercury resistance associated with the sex factor (FP). Genet. Res. 16:179–184.

Mercer, A.A. 1979. Transformation in *Pseudomonas aeruginosa*. PhD. thesis, University of Otago, New Zealand.

Mercer, A.A., and J.S. Loutit. 1978. Transformation in *Pseudomonas aeruginosa* by DNA of the conjugative plasmid FP2. Proc. Univ. Otago Med. Sch. 56:56−57.

Mercer, A.A., and J.S. Loutit. 1979. Transformation and transfection of *Pseudomonas aeruginosa*: effects of metal ions. J. Bacteriol. 140:37−42.

Olsen, R.H., G. Debusschen, and W.R. McCombie. 1982. Development of broad host-range vectors and gene banks: self-cloning of the *Pseudomonas aeruginosa* PAO chromosome. J. Bacteriol. 150:60−69.

Potter, A.A. 1981. DNA synthesis in *Pseudomonas aeruginosa*. Ph.D. thesis, University of Otago, New Zealand.

Potter, A.A., and J.S. Loutit. 1982. Exonuclease activity from *Pseudomonas aeruginosa* which is missing in phenotypically restrictionless mutants. J. Bacteriol. 151:1204−1209.

Sakaguchi, K. 1982. Vectors for gene cloning in *Pseudomonas* and their application, 31−45. *In* P.H. Hofschneider, and W. Goebel (ed.), Current Topics in Microbiology and Immunology. Springer-Verlag, Berlin.

Wood, D.O., M.F. Hollinger, and M.B. Tindol. 1981. Versatile cloning vector for *Pseudomonas aeruginosa*. J. Bacteriol. 145:1448−1451.

chapter 4

MUTANT CONSTRUCTION BY *IN VITRO* MUTAGENESIS

S. Gillam, M. Zoller, and M. Smith

Classical methods for the mutation and manipulation of organisms and the identification and analysis of mutant derivatives are inefficient and time consuming because they involve the random mutagenesis of total genomic DNA. The proportion of mutants that are of the desired phenotype is usually small and is obtained only after exhaustive screening of very large numbers of isolates. Moreover, such methods do not allow the isolation of derivatives containing mutations that are lethal (no assay is available) or which do not produce a marked change in phenotype(s). Classic methods therefore permit the analysis of only certain aspects of the function of any given gene and only certain genes in any given genome.

With the technological triad of restriction endonuclease dissection of DNA, recombinant DNA molecular cloning, and rapid DNA sequence determination, it is now possible to precisely modify DNA sequences *in vitro*, prior to studies carried out either *in vitro* or *in vivo* on the genetic functions of the modified DNA. An important adjunct to most *in vitro* mutant constructions is the cloning of the target DNA in a host cell where the mutated DNA is nonessential. A variety of strategies for mutant construction have been developed, ranging from production of random deletions to specific changes of individual nucleotides. There is no uniquely best method for *in vitro* mutant construction. The method of choice for a given DNA depends on the questions being asked and the information already available on the DNA sequence.

A number of strategies for construction of deletions, insertions, and point mutations *in vitro* will be presented in the following sections. It is clear that a

157

spectrum of methods is essential for a systematic and complete study on DNA function. It is exciting to know that, with the present methods, one can ask and answer very specific questions about the function of any nucleotide in a DNA sequence.

I. Deletion Mutants

Cleaving small genomes with restriction endonucleases and ligating the DNA after removal of specific fragments provided the first and most simple route to *in vitro* construction of mutants (Lai and Nathans, 1974). Many procedures have been developed for deleting DNA at preselected sites. All involve an initial directed incision of a circular DNA duplex (generally in the form of a recombinant bacterial plasmid): (a) at a unique site, by digestion with a restriction endonuclease that cleaves the genome once; or, (b) at one of several specific sites, by partial digestion with a restriction endonuclease that cuts the genome at multiple locations. The termini generated by any single restriction endonuclease that produces cohesive ends, and all that produce fully base-paired ends, can be ligated *in vitro* with T4 DNA ligase without further modification. When the termini are incompatible, they can be enzymatically removed by S1 nuclease, or, in the case of 5'-end tails, filled-in with DNA polymerase prior to blunt end ligation with T4 DNA ligase.

Several methods for generating deletion mutants independent of restriction endonuclease cleavages have been developed. In the first procedure, a nonspecific nuclease is used to produce unit length linear DNA from circular DNA by limited digestion with pancreatic DNase I in the presence of $MnCl_2$ (Shenk *et al.*, 1976). The linear DNA molecules are shortened and recircularized before transformation. The second procedure for random linearization of a circular duplex involves the use of ethidium bromide to limit DNase digestion to a single nick (Cole *et al.*, 1977). The nick can then be extended into a small gap by limited exonucleolytic digestion (Shortle and Nathans, 1978) and the circle opened at the gap by S1 nucleases (Pipas *et al.*, 1980). Re-ligation of the deleted linear molecules generates a family of deletion mutants (Peden *et al.*, 1980). In the third method, a single-stranded segment of DNA is used to relax the superhelical DNA by formation of a displacement loop (D-loop). The loop is then endonucleolytically degraded using S1 endonuclease. After recircularization with DNA ligase a family of short deletions is obtained, all derived from the region of the D-loop (Green and Tibbetts, 1980). In the fourth procedure, a synthetic oligodeoxyribonucleotide corresponding to the desired sequence alteration is used to construct a deletion mutant (Wallace *et al.*, 1980).

These *in vitro* deletion procedures have been widely used to examine the functions of small genomes, such as that of SV40 (Lai and Nathans, 1974; Shenk *et al.*, 1976; Covey *et al.*, 1976), to identify essential elements for transposition of the ampicillin resistance transposon Tn3 (Heffron *et al.*, 1977), to define the promoter region of the *Xenopus* 5S RNA gene (Sakonju *et al.*, 1980), and to identify the different functional regions of the promoter of the iso-1-cytochrome C yeast gene (Faye *et al.*, 1981).

METHOD 1: GENERATING SIMPLE DELETIONS

A. EXCISIONAL DELETION

In this procedure the circular recombinant DNA is digested with a restriction endonuclease at two sites within the plasmid DNA molecule, followed by recircularization and recloning.

PROCEDURE	COMMENTS
1. Digest the DNA with a restriction endonuclease (see Chapter 1, Section IX, p. 35).	**1.** As mentioned previously (see Chapter 1, Section IX, p. 37), each restriction enzyme has a set of optimal reaction conditions (temperature of incubation and the composition of the buffer). In setting up the digestion, the concentration of DNA should be about $0.2-1$ μg in 10 μL of reaction mixture. In general, 1 unit of restriction enzyme is used to digest 1 μg of DNA, under appropriate reaction conditions, in 60 minutes. Digestion for longer periods of time or with excess enzyme does not cause problems unless there is contamination with DNase or exonuclease.
2. Ligate the DNA molecule with T4 DNA ligase as described in Chapter 1, Section X, p. 50.	**2.** To minimize recircularization of the original molecule, the digested DNA may be run on agarose gels (Chapter 1, Section V, p. 18) and the appropriate fragment extracted from the gel (Chapter 1, Section VIII, p. 30).

B. EXONUCLEOLYTIC DELETION

This method uses the exonuclease activity of *Bal*31 to remove nucleotides from both 5′- and 3′-termini of double-stranded DNA (Lau and Gray, 1979). Under suitable conditions, a linear DNA molecule can be digested from both ends in a controlled fashion by inactivating the enzymatic activity with EGTA (ethylene glycol-bis[β-aminoethylether-N,N,N′,N′-tetraacetic acid]), as *Bal*31 is absolutely dependent on Ca^{++} and can be completely inactivated by EGTA, which chelates Ca^{++} specifically.

PROCEDURE	COMMENTS
1. Produce a linear DNA molecule by cleaving with a given restriction endonuclease.	**1.** Alternatively, where no restriction endonuclease sites are available in the recombinant DNA molecule, a linear DNA fragment can be obtained by DNase I digestion (Shenk *et al.*, 1976) or sonication with a sonic probe at low power setting with four bursts of 5 seconds each. Allow 1 minute of cooling between bursts.
2. Add *Bal*31 to the mixture.	**2.** In general, a reaction with *Bal*31 is set up and samples are withdrawn at different times into EGTA. The extent of digestion by *Bal*31 can be analyzed by gel electrophoresis.
a. Incubate the DNA sample (usually at a concentration of 20 μg/mL) with *Bal*31 (87 units/mL) at 30°C in *Bal*31 buffer.	***a.*** **Bal31 Buffer:** 20 mM Tris-HCl (pH 8.1) 12.5 mM $MgSO_4$ 12.5 mM $CaCl_2$ 0.6M NaCl 1 mM EDTA
b. Withdraw 5 μL sample at 0, 15, 30, 45, and 60 minutes and quench the reaction by the addition of 0.1 volume of 0.5M EGTA.	

PROCEDURE

c. Analyze the extent of digestion by gel electrophoresis.

3. The recessed 3'-ends left after treatment of double-stranded DNA with *Bal*31 must be *filled-in* as follows:

a. Extract DNA with phenol and precipitate with ethanol (Chapter 1, Section I, p. 1).

b. Dissolve the DNA pellet in 20 μL of Klenow buffer.

c. Add 1 μL of 2.5 mM stock solution of four dNTPs.

d. Mix and add 1 unit of DNA polymerase I (Klenow fragment).

e. Incubate at 25°C for 30 minutes.

f. Stop the reaction by adding 1 μL of 0.5M EDTA.

4. **Blunt-end Ligation:**

a. Dissolve the DNA fragment in 20 μL of Ligation buffer.

b. Add 5−10 Weiss units of T4 DNA ligase.

c. Incubate at 14°C for 12−16 hours.

5. Transform the appropriate cells. (Chapter 1, Section XIV, p. 81).

COMMENTS

3. Because removal of the 3' and 5' termini of duplex DNA is not synchronous, only a fraction of the DNA molecules in the reaction mixture at any one time will have blunt ends that are suitable for ligation. *Bal*31-treated DNA should be repaired in an end-filling reaction using the Klenow fragment of *E. coli* DNA polymerase I (see Chapter 1, Section XII, p. 71).

b. **Klenow Buffer:**

    50 mM Tris-HCl, pH 7.2
    10 mM $MgCl_2$
    0.1 mM dithiothreitol
    50 mM NaCl

c. **Stock Solution of Four dNTPs:**

Mix the following.

    10 μL 10 mM dATP
    10 μL 10 mM dTTP
    10 μL 10 mM dCTP
    10 μL 10 mM dGTP

4. Ligation of blunt-ended DNA requires a high concentration of enzyme and a high concentration of DNA ends (greater than 1 μM). Therefore, very large amounts of the restriction fragment to be cloned are needed.

a. **Ligation Buffer:**

    20 mM Tris-HCl, pH 7.6
    10 mM $MgCl_2$
    10 mM dithiothreitol
    0.6 mM ATP

b. One Weiss unit is the enzyme activity, which exchanges at 37°C 1 nmol [32]PPi into Norite absorbable material within 20 minutes.

5. Factors that affect the probability of genetic transformation of *E. coli* by plasmids have been evaluated by Hanahan (1983). A set of conditions is described

under which about one in every four hundred plasmid molecules produces a transformed cell. These conditions include cell growth in medium containing elevated levels of Mg^{++}, and incubation of the cells at 0°C in a solution of Mn^{++}, Ca^{++}, Rb^+ or K^+, dimethylsulfoxide, dithiothreitol, and hexamine cobalt.

METHOD 2:   GENERATING FINE DELETIONS

An example of generating a fine deletion is shown in Figure 4.1. At first, a specific fragment *(ab)* is obtained from a given point by the digestion of a defined DNA fragment in a recombinant DNA molecule with restriction endonucleases (1). The fragment is purified and digested with an exonuclease (2), such as exonuclease III, which degrades DNA in the 3'→5' direction. After the approximate desired deletion size is obtained, the single-stranded regions of the DNA strand are removed by S1 nuclease digestion (3). The mutated DNA molecule must then be inserted into the gene from which it was derived. This is usually done by the addition, using blunt-end ligation, of synthetic linkers (4) which contain a useful restriction site. If deletions from only one end are required, the DNA segment is cleaved in the central region of the molecule with restriction enzymes to separate the two fragments (5). The desired segment is then reinserted into the gene and cloned into *E. coli* (6). In this way, a number of clones will be generated, each lacking different discrete segments of the gene region under investigation. DNA sequence analysis is used to determine precisely the end points of the deletions.

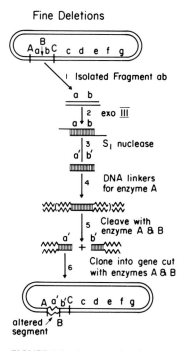

Fine Deletions

FIGURE 4.1   An example of methods used in construction of fine deletion mutants. A, B, and C are used to designate specific restriction enzymes; a−g are used to represent the specific regions of DNA in the recombinant molecule.

PROCEDURE

1. Isolate the DNA fragment as follows:

a. Excise a specific fragment (e.g., Fig. 4.1, fragment ab) from a recombinant plasmid by digestion with the appropriate restriction endonuclease in appropriate buffer (see Chapter 1, Section IX, p. 37).

b. Separate the desired fragment by electrophoresis in either agarose or acrylamide gels in TBE buffer.

c. Excise the DNA fragment from the gel by eluting electrophoretically into 0.05× TBE (see Chapter 1, Section VIII, p. 32).

d. Precipitate the eluted DNA by adding 0.1 volume of 3M sodium acetate (pH 5.5) and 2.5 volumes of ethanol.

2. Digestion with **EXONUCLEASE III**

a. Add 0.1 mL of Exonuclease buffer to the vacuum-dried DNA pellet (about 10 pmoles).

b. Add 100 units of exonuclease III.

c. Incubate at 27°C for 10 minutes.

d. Add EDTA to 10 mM.

e. Extract DNA with phenol, followed by ether and then precipitate with ethanol (Chapter 1, p. 1, Section I).

3. Treatment with **S1 NUCLEASE**

a. Dissolve the DNA pellet from the previous ethanol precipitation in 20 μL of S1 buffer.

b. Add S1 nuclease (the amount of enzyme titrated quantitatively to remove overhanging ends).

COMMENTS

1. Construction of deletion mutants by removal of specific fragments is limited by the random distribution of suitable pairs of restriction endonuclease cleavage sites. Consequently, a number of methods have been developed which require only one restriction endonuclease cleavage or which are completely independent of such cleavage (Shortle and Nathans, 1978; Shenk *et al.*, 1976).

b. **TBE Buffer:**

90 mM Tris-borate, pH 8.3
2.5 mM EDTA

a. **Exonuclease Buffer:**

70 mM Tris-HCl, pH 8.0
4 mM $MgCl_2$
5 mM dithiothreitol

3. S1 nuclease removes single-stranded tails from DNA fragments to produce blunt ends. However, duplex DNA is digested by S1 if it is exposed to very large amounts of the enzyme.

The amount of nuclease S1 required is determined by carrying out a set of pilot-scale reactions, each containing the same amount of DNA fragments but varying in the amount of S1 nuclease (0, 1, 2, 4, or 6 units of nuclease in 20 μl of reaction buffer).

a. **S1 Buffer:**

50 mM sodium acetate, pH 4.5
0.15M NaCl
0.5 mM $ZnSO_4$

c. Incubate at 25°C for 30 minutes.

d. Add 2 µL 0.1M EDTA.

e. Extract the DNA with phenol, ether, and precipitate with ethanol.

4. Ligate synthetic linkers to deleted fragments as follows:

4. In order to facilitate tracing DNA through the subsequent steps (the separation of the DNA fragments from the linkers by chromatography on Sepharose CL-4B or Biogel A 0.5M), synthetic linkers are radioactively labelled using T4 polynucleotide kinase before ligation (see Section III, p. 183).

a. Radioactively label the linkers using T4 polynucleotide kinase (Reverse Reaction; see Chapter 1, p. 76, Section XII).

b. To a siliconized tube add:

5 µL of the deleted fragments (e.g., Fig. 1, fragment ab)

4 µL of the labeled linker (20× molar excess over fragment ab)

1 µL 10× ligation buffer, (p. 160).

c. Add 3 units of T4 DNA ligase.

d. Incubate at 14°C overnight.

e. Stop the reaction by the addition of 1 µL 0.1M EDTA, followed by phenol extraction.

f. Concentrate the DNA by precipitating with ethanol.

5. Cleave with restriction enzymes A and B (see Fig. 4.1).

5. Because the molar amount of the linkers is so large, a large amount of restriction enzyme is required to achieve a complete digestion. The digested linkers can also join to the vector, leading to a high background of nonrecombinant plasmids. Therefore, it is important to separate the DNA fragment from the linkers before and after endonuclease digestion. This can be accomplished by gel filtration chromatography.

a. Digest the DNA with a high concentration of restriction enzyme A (100 units) and enzyme B (2 units) in 50 µL using appropriate digestion conditions.

b. Incubate for 60 minutes at the appropriate temperature.

c. Load onto a 5 mL column of Biogel A-0.5M or Sepharose CL-4B column equilibrated with 0.3M sodium acetate (pH 6.5).

d. Elute the column with 0.3M sodium acetate. Collect twelve 0.3 mL fractions and analyze 50 µL of each fraction by agarose gel electrophoresis.

e. Pool the fractions containing the DNA fragments and precipitate the DNA with ethanol.

6. Using the ligation procedure described in Method 1B, p. 160, clone the fragment into a cloning vector.

7. Transform into the appropriate cells (Chapter 1, p. 81, Section XIV).

II. Insertion Mutants

The strategies for producing the deletion mutants described above provide a powerful set of tools for the construction of mutants which disrupt specific DNA function. A related strategy involves the introduction of a DNA insert into the genome to produce defective mutants. Insertions of DNA not only interrupt a DNA sequence at a desired position, but also may provide a new restriction target or a new functional element. The end to end ligation of DNA fragments with complementary or fully base paired termini has been widely used to insert DNA segments into vector molecules to construct novel combinations of genes and regulatory signals (Maniatis, 1980; Bechman *et al.*, 1976; Mulligan and Berg, 1980; Post *et al.*, 1981). The attraction of this approach lies in the fact that the inserted fragment can contain a specific restriction endonuclease cleavage site. This facilitates physical mapping of the mutant sites relative to a marker restriction endonuclease site. The physical map can be correlated with the corresponding genetic map.

The *in vitro* insertion of DNA fragments into target molecules may be site-specific or random. The simplest method for the site-specific introduction of insertions is at the cleavage site of specific restriction endonucleases that generate fragments with 5′-cohesive single-stranded termini (see Figure 4.2A). This approach has been used to produce a fine structure map of RSF1050, a *Col*EI derivative (Backman *et al.*, 1978). In addition, by using this method, Boeke (1981) constructed one and two amino acid codon insertions in the circular genome of bacteriophage f1. Full length linear molecules were produced by limited digestion of f1 replicative form DNA with *Hin*fI, a restriction endonuclease that leaves a three base 5′ extension. Repair with DNA polymerase I and deoxynucleoside triphosphates and recircularization with DNA ligase resulted in a three base duplication of the sequence at the sites of initial cleavage. Since one codon was added to a protein-coding sequence, no frameshift mutations would be generated. Similarly, enzymatic repair of linear molecules produced by digestion with *Hpa*II (which leaves a two base extension) and blunt-end insertion of a tetranucleotide resulted in the formation of a set of two codon insertions.

The above insertion procedures are site-specific. They can, however, be used to mutagenize a genome at random locations if the target DNA is linearized not with a restriction endonuclease but by DNase I in the presence of Mn^{++} (Shenk *et al.*, 1976). The blunt-end ligation of linkers to randomly linearized hybrid molecules is a highly useful method for introducing short insertions at random locations in small genomes. A large number of mutants may be isolated at defined sites that are readily localized on the physical map through restriction endonuclease mapping. This method has been used to map the ST toxin gene of *E. coli* (So and McCarthy, 1980) and to map the genetic functions of the mating type loci, *MATa* and *MATα* of *S. cerevisiae* (Tatchell *et al.*, 1981). The use of linkers to generate readily mapped, random insertions in cloned DNA fragments is a very powerful method for the functional analysis of any cloned gene or operon; it is expected to find widespread use in the future.

An alternative procedure for random linearization of a circular duplex involves the use of ethidium bromide to limit pancreatic DNase digestion to a single nick. The nick can then be extended into a small gap by limited exonucleolytic digestion and the circle opened at the gap by S1 nuclease.

METHOD 1: SITE-SPECIFIC INSERTIONS

This procedure (Figure 4.2A) involves linearization of the recombinant DNA molecules with the appropriate restriction endonuclease (1, EcoRI) filling-in the single-stranded termini with polymerase I (2), recircularization of the molecules by blunt-end ligation with T4 DNA polymerase (3) and reintroduction of the molecules into the host cell system.

PROCEDURE

1. Digest the molecule with a specific restriction endonuclease (EcoRI) (see Chapter 1, Section IX, p. 35).

2. Produce blunt ends by filling-in recessive 3′-ends (see Chapter 1, Section XII, p. 70).

a. To a siliconized microfuge tube add:

1 μg DNA
2 μL 1 μM dTTP and dATP
2 μL 10× Klenow buffer
H$_2$O to 20 μL

b. Add 1 unit Klenow fragment of DNA polymerase.

c. Incubate at 25°C for 15−30 minutes.

d. Heat at 70°C for 5 minutes to inactivate the enzyme.

3. Add 2 μL of 5 mM ATP to the reaction mixture and ligate as described previously.

4. Transform the appropriate cells. (Chapter 1, Section XIV, p. 81).

COMMENTS

1. The recognition sequence for EcoRI is

$$\downarrow$$
5′-G-A-A-T-T-C-3′   5′-G-3′
3′-C-T-T-A-A-G-5′ $\longrightarrow$ EcoRI   3′-C-T-T-A-A-5′
$$\uparrow$$

2. To fill-in recessed 3′-ends created by cleavage with DNA by EcoRI, only dATP and dTTP are needed in the reaction.

a. **10× Klenow Buffer:**

500 mM Tris-HCl, pH 7.5
100 mM MgCl$_2$
10 mM DTT
500 mM NaCl

3. There is no need to purify the blunt-ended DNA since the end filling and ligation reactions can be carried out sequentially in the same reaction mixture (Section I, p. 159 Method 1B).

METHOD 2: RANDOM INSERTIONS

For random insertion by using chemically synthesized linkers (Figure 4.2B), full length linear DNA molecules of a plasmid are purified following random limited pancreatic DNase I digestion in the presence of Mn^{++}(1). After the termini are made blunt by repair with DNA polymerase I (2), a synthetic linker containing a specific restriction endonuclease site (EcoRI) is joined to the ends by using T4 DNA ligase (3). Excess linker is removed and the ends rendered cohesive by appropriate restriction enzyme digestion (4). Following circularization with E. coli DNA ligase (5), which joins molecules with cohesive ends only, and transformation, the plasmids contain new inserted cleavage sites. This method yields a new circular plasmid DNA into which a single new restriction site (EcoRI) has been randomly inserted.

A. Site-specific Insertion

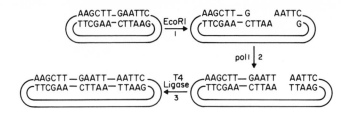

B. Random Insertion

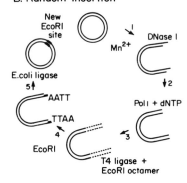

FIGURE 4.2 Schemes for the construction of insertion mutants (A) containing a short oligodeoxyribonucleotide at a specific site, (B) random insertion of a new restriction site in the plasmid.

PROCEDURE

1. Treat with DNase I as follows:

a. To a siliconized tube add:

plasmid DNA (100 μg)
20 μL 10× DNase buffer
H₂O to 200 μL

b. Mix and add 0.3 pg of DNase I.

c. Incubate at 25°C for 20 minutes.

d. Extract the DNA with phenol, ether and concentrate by ethanol precipitation.

COMMENTS

1. DNase I is an endonuclease that hydrolyzes double-stranded or single-stranded DNA. In the presence of Mg^{++}, DNase I attacks each strand of DNA independently and the sites of cleavage are distributed in a statistically random fashion. However, in the presence of Mn^{++}, it cleaves both strands of DNA at approximately the same site to yield fragments of DNA that are blunt ended or have protruding termini only one to two nucleotides in length (Melgar and Goldthwaite, 1968).

a. **10× DNase Buffer:**

200 mM Tris-HCl, pH 7.5
15 mM MnCl₂
1000 μg/mL bovine serum albumin

b. Make a serial dilution of a small quantity of a stock solution of DNase I (1 mg/mL). Add 1 μL of diluted DNase I (0.3 ng/mL). $1 \mu g = 10^6$ pg.

PROCEDURE	**COMMENTS**

2. Fill-in recessed 3'-ends using DNA polymerase I (Klenow fragment).

2. In order to obtain flush-ended molecules, linear unit length plasmid DNA (25 µg) is treated with DNA polymerase I (Klenow fragment) in the presence of four dNTPs (see Section I, Method 1B, p. 159).

3. Ligate an *Eco*RI octamer linker.

3. See Section I, Method 2 (p. 161).

4. Cleave with *Eco*RI restriction enzyme.

4. See Section I, Method 2 (p. 161).

5. Ligate with *E. coli* ligase.

5. The *E. coli* ligase, which does not have flush-end ligation activity, gives a lower background of circular molecules lacking an added restriction site.

a. To a siliconized tube add:

DNA fragments (1 µg)
10 µL 10× *E. coli* ligase buffer
H₂O to 100 µL

a. **10× *E. coli* Ligase Buffer:**

200 mM Tris-HCl, pH 7.5
100 mM MgCl₂
  1 mM NAD
500 (µg/mL) bovine serum albumin

b. Mix and add 0.5 unit of *E. coli* ligase

b. *E. coli* ligase requires NAD for its activity (present in *E. coli* ligase buffer).

c. Incubate at 14°C for 12–16 hours.

d. Transform the appropriate cells (Chapter 1, Section XIV, p. 81).

III. Construction of Point Mutants Introduced by Synthetic Oligodeoxyribonucleotides

Deletion and insertion mutants define only the location and boundaries of genetic functions. In order to define the role of specific nucleotides, it is necessary to construct point mutants within a defined DNA segment. A wide variety of *in vitro* mutagenesis procedures which produce site-specific mutants have been developed. These include chemical modification of a specific DNA fragment followed by its reintegration into genomic DNA (Borrias *et al.*, 1976), mutations induced by specific chemically modified RNA (Salganik *et al.*, 1980), chemical modification of a specific single-stranded gap in duplex DNA (Shortle *et al.*, 1979) and enzymatic incorporation at a specific site of a mutation-inducing nucleotide analog (Weissman *et al.*, 1979). However, these methods do not allow for the production of all types of point mutation (transitions, transversions, insertions, and deletions), and full characterization of mutant DNA requires biological cloning and DNA sequence determination of many clones.

With the recent development of solid phase methods for the chemical synthesis of oligodeoxyribonucleotides (Matteucci and Coruthers, 1981; Itakura and Riggs, 1980; Adams *et al.*, 1983) the most precise and versatile route to study gene functions is to introduce a specific mutation in DNA through the use of short synthetic oligodeoxyribonucleotides as site-specific mutagens. Important features of this method include its high efficiency and the precision with which a desired change is determined by the synthetic oligodeoxyribonucleotide of defined sequence. This methodology has been successfully applied to recombinant filamentous phage with vectors derived from fd (Wasylyk *et al.*, 1980) and M13 (Simons *et al.*, 1982; Miyada *et al.*, 1982; Temple *et al.*, 1982; Zoller and Smith, 1982; Winter *et al.*, 1982), and recombinant plasmid DNA (Wallace *et al.*, 1980; Wallace *et al.*, 1981).

Ideally, a mutagenic oligodeoxyribonucleotide will form a stable and unique duplex structure with a complementary region of the genome under study. In the interaction of oligonucleotides and complementary oligonucleotides, the major structural determinants of duplex stability are base composition, nucleotide mismatches, and the length of the oligonucleotide (Astell *et al.*, 1973; Gillam *et al.*, 1975). Oligodeoxyribonucleotides ranging in length from 8–18 nucleotides have been used to direct mutagenesis. Oligonucleotides of such lengths can be synthesized either enzymatically (Gillam and Smith, 1980) or chemically (Matteuci and Caruthers, 1981; Itakura and Riggs, 1980; Adams *et al.*, 1983). The enzymatic procedures produce relatively low yields of final product and proceed by slow empirical reactions. Currently, chemical methods based on solid phase chemistries are the most widely used. In addition, the cost of commercially produced oligonucleotides has dropped significantly with the proliferation of companies offering custom synthesis. Furthermore, automated machines which synthesize oligodeoxyribonucleotides can be purchased thereby making these defined sequences available to all molecular biologists.

Success in producing a specific mutation is greatly dependent on the mutagenic oligodeoxyribonucleotide selected. The mismatched nucleotide residue in the oligodeoxyribonucleotide primer must be protected from editing out by the 5′- and 3′-exonuclease activities of the *E. coli* DNA polymerase I. The 5′-exonuclease activity can be eliminated by using the large fragment of *E. coli* DNA polymerase I (Klenow and Henningsen, 1970). The 3′-exonuclease activity, intrinsic to the polymerase activity of *E. coli* DNA polymerase I, may also edit out the mismatched sequence of the mutagenic oligonucleotides. However, this activity can be overcome by using an oligodeoxyribonucleotide with more than one base residue beyond the 3′-end of the mismatched base. If the mutagenic oligonucleotide is to be used as a probe to screen for the constructed mutant, it is desirable to design the mismatch(es) near the middle of the molecule in order to yield the greatest binding differential between a perfectly matched duplex and mismatched duplex.

Initially, oligodeoxyribonucleotide-directed mutagenesis was developed using bacteriophage ϕX174 as a model system (Hutchison *et al.*, 1978; Gillam *et al.*, 1979, 1980). Recently, we have described a simple and highly efficient procedure for oligonucleotide-directed mutagenesis of DNA fragments inserted into M13-derived vectors (Zoller and Smith, 1982). The advantages of using these phage vectors include the simple and rapid isolation of single-stranded template DNA, easy screening for the mutant clones by using a mutagenic oligonucleotide as a probe, and the fact that the mutated DNA fragment cloned in the M13 vector can be nonessential.

The basic strategy for oligodeoxyribonucleotide mutagenesis is described in Figure 4.3. The mutagenic oligodeoxyribonucleotide with a mismatch that directs the mutation is used as a primer for *in vitro* DNA synthesis with single-stranded circular DNA template. Closed circular heteroduplex DNA synthesized *in vitro* by using *E. coli* DNA polymerase I (Klenow, large fragment) and T_4 DNA ligase is enriched by S_1 endonuclease treatments or alkaline sucrose gradient centrifugation. Upon transformation of *E. coli* cells with the *in vitro*-synthesized closed circular DNA, a population of mutant and wild-type molecules is obtained. The constructed mutant may be screened for by its biological phenotype (Hutchison *et al.*, 1978; Gillam *et al.*, 1979), by the detection of induced or deleted restriction endonuclease sites (Wasylyk *et al.*, 1980), by *in vitro* selection with the mutagenic oligodeoxyribonucleotide (Gillam *et al.*, 1980), by *in situ* hybridization with the mutagenic oligodeoxyribonucleotides (Zoller and Smith, 1982; Winter *et al.*, 1982; Wallace *et al.*, 1980; Wallace *et al.*, 1981) or by DNA sequencing (Gillam and Smith, 1983).

In order to show conclusively that the mutagenic oligodeoxyribonucleotide has induced the intended mutation, the changed nucleotide sequence at the target site in the selected mutant must be determined. A suitable DNA fragment or a short synthetic oligodeoxyribonucleotide may be used as primer for sequence determination by the enzymatic terminator method (Sanger *et al.*, 1977).

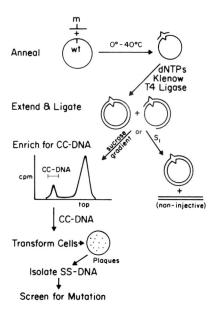

FIGURE 4.3 Strategies for *in vitro* construction of site-specific mutants by using synthetic oligodeoxyribonucleotides as mutagens.

m   = the mutant oligodeoxyribonucleotide with mismatched base (m)
wt   = wild-type DNA template
cc-DNA = closed circular duplex DNA molecules
ss-DNA = single-stranded closed circular DNA molecules

A. PREPARING THE DNA TEMPLATE

METHOD 1:   PREPARING SINGLE-STRANDED CIRCULAR DNA TEMPLATE (M13 SYSTEM)

M13 is a filamentous bacteriophage with a closed circular DNA genome approximately 6500 nucleotides in length (Denhardt *et al.*, 1978). The phage attaches to F-pili of *E. coli* and is able to infect only male bacterial cells. During infection of a suitable *E. coli* host, the infecting single-stranded DNA (+ strand) of the phage is converted and amplified into double-stranded replicative forms (100−200 molecules per cell) which serve as intermediates in the production of progeny single-stranded DNA. These single-stranded DNA molecules are packaged into a protein coat and extruded out of the cell. As the M13 phage does not lyse the host cells, the packaged particles are continually extruded from the infected cells. Extraction of the culture supernatant can yield 5−10 µg single-stranded DNA per mL. The advantage of using M13 as a DNA template is that the mutagenic oligonucleotide can be primed directly on the template for synthesis of heteroduplex DNA and the mutagenic oligonucleotide can be used to screen for the mutant directly on the phage DNA.

The vector M13mp2 contains a portion of the *lac* gene of *E. coli* including operator, promoter, and the N-terminal 145 amino acid residues of β-galactosidase (the so called α-peptide). M13mp2 has a unique *E. coli* cloning site near the N-terminus of the α-peptide of β-galactosidase. M13mp7 is derived from M13mp2 but contains a synthetic oligonucleotide with an array of restriction sites (*Eco*RI, *Bam*HI, *Sal*I, *Pst*I, *Acc*I, *Hin*dIII) inserted in the original *Eco*RI site. More recently, further M13 derivatives, M13mp8, M13mp9, M13mp10, and M13mp11 have become available (Messing, 1983).

PROCEDURE

1. Grow overnight cultures of JM101 by inoculating 5 mL of supplemented M9 media (see Section E, p. 178) with a single colony of JM101 from a minimal plate (+ glucose).

2. Add 0.1 mL of the overnight culture to 10 mL of 2× YT, shake at 37°C water bath for 60 minutes.

3. Using a sterile toothpick, pick a plaque of M13 and inoculate into the 10 mL 2×YT culture.

4. Shake at 37°C for 5–6 hours.

5. Centrifuge 5 minutes at 5,000 rpm.

6. Pour the supernatant into another centrifuge tube.

7. Add 2.5 mL of solution containing 2.5M NaCl and 20% PEG (M.W. 6000).

8. Mix and leave at 4°C for at least 2 hours.

9. Centrifuge at 10,000 g for 5 minutes and remove the supernatant.

10. Suspend the phage pellet in 0.5 mL 10 mM Tris, pH 7.4, 0.1 mM EDTA.

11. Extract the phage suspension with an equal volume of phenol and chloroform mixture (1:1).

12. Remove traces of phenol by extracting with diethyl ether.

13. Concentrate phage DNA by ethanol precipitation.

COMMENTS

1. JM101 host strain requires proline and is derived from *E. coli* K12 (p. 109).

By growing the host on minimal media (+ glucose) the plasmid proline gene can be used as selection against losing the episome. It is important to maintain the plasmid because M13 is a male-specific phage and thus will not infect female cells.

2. 2× YT: (100 mL)

1.6 g tryptone
1.0 g yeast extract
0.5 g NaCl

3. See Chapter 1, Section XVII, p. 109 for propagating M13.

7. M13 phage particles are like T4 phage or other virus particles. They can be precipitated by polyethylene glycol in the presence of salt.

9. Carefully remove all the remaining traces of PEG supernatant with a Pasteur pipette.

METHOD 2: PREPARING A CIRCULAR SINGLE-STRANDED DNA TEMPLATE FROM DOUBLE-STRANDED RECOMBINANT PLASMIDS

This procedure is based on a method described by Wallace *et al.*, (1981).

PROCEDURE

1. Randomly nick the plasmid (i.e., pBR322 DNA) by incubating 600 μg of plasmid DNA in 1 mL of Hin buffer in the presence of ethidium bromide (100 μg/mL) and pancreatic DNase I (150 mg/mL).

2. Analyze the extent of nicking by agarose gel electrophoresis (Chapter 1, Section V, p. 18).

3. Incubate 10 μg of nicked pBR322 in 0.1 mL of exonuclease buffer (p. 162) with 60 units of exonuclease III for 90 minutes at 37°C.

4. Add 10 μL of 0.1M EDTA. Extract the reaction mixture with phenol/chloroform and, after the addition of sodium acetate to 0.3M, recover the single-stranded circular DNA by ethanol precipitation.

COMMENTS

1. 10× HIN BUFFER:

66 mM Tris-HCl, pH 7.4
66 mM $MgCl_2$
0.5M NaCl
10 mM dithiothreitol

With a single-stranded DNA, a mutagenic oligodeoxyribonucleotide is used directly to prime the synthesis of DNA on a circular DNA template. Because the mutagenic oligodeoxyribonucleotide has to be covalently integrated into DNA at both ends (Gillam *et al.*, 1975), a circular DNA is the most convenient substrate. With plasmids, such as pBR322, the single-stranded circular template can be prepared by nicking the supercoiled DNA with restriction endonuclease or pancreatic DNase I in the presence of ethidium bromide (Wallace *et al.*, 1981) to generate a single nick in the DNA. This nicked DNA is a substrate for exonuclease III and digestion produces single-stranded circles.

2. Nicked plasmids move slower than the covalently closed circular DNA in agarose gels. Approximately 80–90% of the plasmid is nicked using this procedure.

B. DETERMINATION OF THE SPECIFICITY OF THE MUTAGENIC OLIGODEOXYRIBONUCLEOTIDE

METHOD 1: PULSE-CHASING PRIMING

PROCEDURE

1. Add to a siliconized 0.5 mL Eppendorf tube:

COMMENTS

1. Success in producing a specific mutation is greatly dependent on the mutagenic oligonucleotide selected.

PROCEDURE

5–20 pmol mutagenic oligonucleotide
0.3 pmol single-stranded DNA template
1 μL 10× Klenow buffer
H₂O to 5 μL

2. Heat at 60°C for 5 minutes. Cool at room temperature for 5 minutes.

3. Add to the annealed DNA mixture:

1 μL each of 0.5 mM dCTP, dGTP, and dTTP
1 μL 0.1 mM dATP
5 μCi ³²P-α-dATP (3,000 Ci/mmol)

4. Add H₂O to total volume of 10 μL. Mix.

5. Add 1 unit of DNA polymerase I (Klenow fragment). Incubate at room temperature for 1 minute.

6. Add 1 μL 0.5 mM dATP. Incubate for 5 minutes at room temperature.

7. Stop the reaction by adding 2 μL of 0.25M EDTA and 10 μL of phenol. Vortex, remove the aqueous phase after centrifugation in an Eppendorf microfuge.

8. Extract the aqueous phase two times with ether to remove phenol. Remove deoxyribonucleoside-5′-triphosphates by passage through a 1 mL Sephadex G-100 column equilibrated and eluted with 2mM Tris-HCl, pH 7.8 and 0.05 mM EDTA. Collect the first radioactive peak from the column.

9. Concentrate the eluted DNA by freeze drying in a desiccator under vacuum, then dissolve the pellet in 30 μL H₂O.

COMMENTS

The specificity of the oligonucleotide can be determined by oligonucleotide-extended synthesis with ³²P-labeled substrate as a pulse, followed by a chase of nonradioactive substrate. Subsequent restriction endonuclease cleavage of the double-stranded product followed by gel electrophoresis gives a measure of the distance between the priming and restriction sites. A single, labeled fragment should be produced if priming occurs at a unique site.

10× Klenow Buffer:

0.5M Tris-HCl, pH 7.5 at 23°C
0.1M MgCl₂
0.5M NaCl
0.01M DTT

The concentration of oligonucleotide can be calculated from the absorbency of the solution by using the average molar absorbency index of 10×10^3 for each nucleotide.

2. The purpose of heating the solution at 60°C for 5 minutes is to denature the DNA template, then anneal at room temperature with oligonucleotide.

5. It is important to label only a short segment of the synthesized DNA. Extended labeling of the DNA fragment will produce more than one DNA fragment after endonuclease digestion as analyzed in denaturing polyacrylamide gels.

6. The concentration of dATP is increased by the addition of cold dATP in this chase reaction.

8. The first radioactive peak eluted from the column contains the ³²P-labeled DNA fragments, and the second radioactive peak contains the ³²P-labeled dATP. Usually, the first radioactive peak elutes at tubes number 3 and 4 (0.2 mL/tube).

10. Add 1 μL of the appropriate 10× buffer and 1 unit of the appropriate restriction enzyme to 9 μL of DNA sample.

11. Incubate the reaction mixture at the appropriate temperature for 60 minutes.

12. Add 10 μL of formamide-dye-EDTA solution. Boil 3 minutes and chill on ice immediately.

12. **Formamide-dye-EDTA Solution:** (for denaturing gels)

> 0.02% bromphenol blue
> 0.02% xylene cyanol FF
> 0.025M EDTA
> in 90% formamide (deionized)

13. Electrophorese the sample on a 7M urea−5% polyacrylamide gel (see Chapter 1, Section XV, p. 89).

13. Use ^{32}P-labeled DNA fragments of known nucleotide length as size markers.

14. After electrophoresis, autoradiograph the gel as indicated in Chapter 1, Section XIII, p. 79.

METHOD 2:   DNA SEQUENCING BY THE CHAIN TERMINATION PROCEDURE

A more precise test for the target specificity of a mutagenic oligonucleotide is to use it as a primer for DNA sequence determination. Specific priming at the desired site results in an unambiguous sequence pattern for the adjacent DNA. If priming is nonspecific, the oligonucleotide will have annealed to a variety of sites (not a single site) and no specific sequence will be observed.

A primer annealed to a single-stranded template can be extended in the 5′→3′ direction by DNA polymerase I (Klenow fragment) in the presence of deoxynucleoside triphosphates (dNTPs). Klenow polymerase will also specifically incorporate dideoxynucleoside triphosphates (ddNTPs) in place of their dNTP counterparts, but the lack of a 3′-hydroxyl group on the ribose moiety of the ddNTP prevents further polymerization. Thus the ddNTPs act as specific chain terminators and can be used to determine the nucleotide sequence beyond the 3′-end of the primer (Sanger *et al.*, 1977). In each of four separate sequencing reactions, the primer is extended in the presence of a different ddNTP (ddTTP, ddCTP, ddGTP, or ddATP) such that there is only partial incorporation of the terminator.

The four sequencing reactions yield four sets of products, each with fixed 5′-end (the primer) and terminating at either T, C, G, or A. If the primer extension is radioactively labeled by including α-^{32}P-dATP in the polymerization, the nucleotide sequence of the template beyond the 3′-end of the primer may be deduced after fractionation of the products on the denaturing polyacrylamide gel and autoradiography of the gel.

PROCEDURE   **COMMENTS**

1. Add to a siliconized tube:

0.1−0.3 pmol recombinant DNA (10−30× molar excess over template DNA)

1 μL 10× Klenow buffer
H$_2$O to 10 μL

2. Mix and seal in a glass capillary tube (0.8 × 100 mm).

3. Boil at 100°C for 3 minutes. Cool quickly on ice.

4. Anneal the template and primer at the appropriate temperature for 5−30 minutes depending on the chain-length of the mutagenic oligonucleotide.

5. Remove the annealed sample from the capillary tube. Add 1 μL of ^{32}P-α-dATP (3,000 Ci/mmole) and 1μL of 0.1 mM dATP.

6. Add 2.5 μL of the above mixture to each of 4 siliconized glass tubes, labeled dC, dT, dA and dG, respectively.

7. Add 2.5 μL of ddC mix to dC-tube, 2.5 μL ddT mix to dT-tube, 2.5 μL ddA mix to dA-tube, and 2.5 μL ddG mix to dG-tube.

2. In order to prevent evaporation during the boiling, the reaction mix is sealed in a capillary tube.

3. The template DNA is denatured by boiling and stays denatured by quickly cooling in ice water.

4. The effects of the chain length of mutagenic oligonucleotides and temperature on the priming efficiency of the DNA template have been studied using the bacteriophage φX174 system (Gillam and Smith, 1979). In general, for oligonucleotides with a chain length of less than 8 nucleotides, the annealing temperature can be carried out at 0°−10°C, 25°−35°C for chain lengths of 8 to 14, and 40°−55°C for chain lengths greater than 15 nucleotides.

7.   10× Hin BUFFER:

66 mM Tris-HCl, pH 7.4
66 mM MgCl$_2$
0.5 M NaCl
10 mM DTT

ddC MIX:

0.1 mM dGTP
0.1 mM dTTP
0.005 mM dCTP
0.25 mM ddCTP
1.5× Hin buffer

ddT MIX:

0.1 mM dGTP
0.1 mM dCTP
0.005 mM dTTP
0.5 mM ddTTP
1.5× Hin buffer

ddA MIX:

0.1 mM dGTP
0.1 mM dTTP
0.1 mM dCTP
0.5 mM ddATP
1.5× Hin buffer

PROCEDURE	COMMENTS

ddG MIX:

0.1 mM dCTP
0.1 mM dTTP
0.005 mM dGTP
0.5 mM ddGTP
1.5× Hin buffer

8. Add 0.25 μL of DNA polymerase I (Klenow, large fragment 0.25–0.5 units) to each tube. Mix. Incubate at room temperature for 15 minutes.

9. Add 0.25 μL of the four dNTP mix (2.5 mM) to each tube. Mix and incubate at room temperature for another 15 minutes.

9. See Section I, Method 1B (p. 160) for four dNTP formula.

After an initial synthesis with carrier-free α-^{32}P-dATP, the prematurely terminated chains are extended by chasing with high concentrations of dNTPs.

10. Add 5 μL of formamide-dye-EDTA mix to each tube.

11. Boil at 100°C for 3 minutes. Cool immediately by placing in ice.

11. Before loading the samples into the gel, the synthesized double-stranded DNA should be denatured by boiling and quickly cooled in ice in formamide-dye solution.

12. Load the sample on a 12% polyacrylamide denaturing gel.

C. *IN VITRO* SYNTHESIS OF CLOSED CIRCULAR HETERODUPLEX DNA

METHOD 1: ANNEALING

PROCEDURE	COMMENTS

1. Add to a 0.5 mL siliconized Eppendorf tube:

0.5 to 1.0 pmol recombinant DNA
mutagenic oligonucleotide (10–30 molar excess over DNA template)
1 μL 10× Klenow buffer
H$_2$O to 10 μL total volume

1. Specific priming of oligodeoxyribonucleotide to the DNA template is the key step in the construction of mutants. Tests of the specificity of priming should be made prior to attempting the biological part of the experiment. A series of priming experiments can easily be carried out at different temperatures as described above.

2. Boil at 100°C for 3 minutes. Cool immediately on ice.

3. Anneal the template and oligonucleotide at the appropriate temperature.

3. See Method 2, Section IIIB in this chapter.

4. Prepare ligation solution in a separate vial and keep on ice.

4. Prepare ligation solution in a 0.5 mL vial before adding to annealed DNA. Keep on ice.

PROCEDURE	COMMENTS

Ligation Solution

    1 μL Solution B
    1 μL 10 mM dCTP
    1 μL 10 mM dTTP
    1 μL 10 mM dGTP
    0.5 μL 0.1 mM dATP
    1 μL 10 mM ribo-ATP
    1.4 μL α-^{32}P-dATP (7 μCi/μL)
    1.5 μL T$_4$ DNA ligase (2g/μL)
    2 μL H$_2$O

5. Add 10 μL of ligation solution to the annealed DNA. Mix.

6. Add 2−5 units DNA polymerase (large fragment).

7. Mix. Incubate for 5 minutes at room temperature.

7. For shorter oligonucleotides (less than 10 nucleotides long), carry out this extension reaction at 0−10°C.

8. Add 1 μL 10 mM dATP. Mix.

8. After an initial synthesis with carrier-free α-^{32}P-dATP, the primed DNA synthesis is extended with cold dATP.

9. Incubate at 15°C for 12−20 hours.

D. ENRICHMENT FOR CLOSED CIRCULAR HETERODUPLEX DNA

Single-stranded phage DNA and gapped duplexes which will transform host cells and produce a high background of transformants, are removed by treatment with S1 endonuclease (Method 1) or by alkaline sucrose density gradient centrifugation (Method 2).

METHOD 1: ENRICHING BY S1 NUCLEASE DIGESTION

PROCEDURE	COMMENTS

1. Mix 2 μL of the ligated reaction product with 2 units of S1 nuclease in a total reaction volume of 20 μL containing 0.28M NaCl, 4.5 mM ZnSO$_4$, 30 mM sodium acetate, pH 4.5.

1. The reason for using 0.28M NaCl instead of 0.15M is to avoid the digestion of mismatched base-pairs by S1 nuclease.

2. Incubate at room temperature for 30 minutes.

3. Stop the reaction by adding 5 μL 1M Tris-HCl, pH 8.1.

3. S1 nuclease is only active between pH 4.5−5.0. Therefore, the reaction can be stopped by changing the pH to 8.1.

METHOD 2: ENRICHING BY ALKALINE SUCROSE GRADIENT CENTRIFUGATION

PROCEDURE

1. Prepare a 5–20% sucrose gradient. After the gradient has formed, prepare the sample as indicated below.

2. Add 20 μL 2N NaOH to the ligated sample. Mix.

3. Incubate at room temperature for 5 minutes.

4. Chill on ice for 1 minute.

5. Place the gradient tube in the SW 50.1 bucket.

6. Apply the sample to the top of the gradient using a disposable micro-pipette.

7. Centrifuge at 37,000 rpm, with the brake off, for 2 hours at 4°C.

8. Following centrifugation, collect aliquots in 1.5 mL Eppendorf tubes by puncturing the bottom of the gradient tube. Collect about 30 fractions with about 5–7 drops per fraction.

9. Determine the amount of ^{32}P-cpm in each fraction by scintillation counting (Cerenkov, see Chapter 1, Section XVII-0, p. 106).

COMMENTS

1. PREPARING ALKALINE SUCROSE GRADIENTS: One 5–20% sucrose gradient is prepared for each sample in a 0.5″ × 2″ centrifugation tube. This is used with a SW 50.1 rotor. A step gradient is made with 1 mL each of the sucrose stock solutions, starting with the 20% solution. Carefully let the solution run down the side of the tube using either a serological or Pasteur pipette. Hold the tip close to the top of the previous layer to avoid mixing. Let the tube sit at 4°C for 6–16 hours to linearize the gradient. Alternatively, leave the gradient at room temperature for 2 hours, then at 4°C for another 2 hours.

Alkaline Sucrose Gradient Stock Solutions (5–20%):

x% sucrose (x = 20, 17.5, 15, 10, or 5)
1M NaCl
0.2M NaOH
2 mM EDTA
Autoclave
Store at 4°C

2. Before the application of the DNA sample to the gradient, gapped duplexes are denatured by alkaline treatment.

6. If the top of the gradient solution is too close to the top edge of the centrifugation tube before adding the sample, withdraw 50–100 μL from the top. The final height of the solution should be within 2–3 mm of the tube edge.

9. The CC-DNA usually migrates to the bottom third of the gradient, whereas the single-stranded circular and linear molecules are found in the top third (see Figure 4.3).

10. Pool the closed circular DNA fraction. Add about 50 μL 1M Tris-citrate (pH 5.0) per 300 μL.

10. The Tris-citrate neutralizes the pH. To ensure that the pH has been lowered to pH 7−8.0, spot a 2 μL aliquot onto pH paper.

11. Measure the amount of ^{32}P-labeled DNA in the pooled CC-DNA fraction by scintillation counting and calculate the %CC-DNA formed.

E. TRANSFECTION AND TRANSFORMATION PROCEDURES

The preparation of "competent" cells for transformation by bacterial DNA or for transfection by viral DNA has always been a largely empirical procedure. In transfection, the spheroplasts are usually prepared by EDTA-lysozyme treatment. CaCl$_2$-treated cells may also be used. Yields of phage progeny are about 25% of those produced from spheroplasts.

METHOD 1: TRANSFECTION

1. Preparation of spheroplasts

1. 3×D Medium

Solution 1

  3 g NH$_4$Cl
  0.3 g MgSO$_4$
  0.3 mL 1M CaCl$_2$
24.0 mL glycerol
15.0 g casamino acids
  3.0 ml 1% (w/v) gelatine
600 mL distilled H$_2$O

Solution 2

0.9 g KH$_2$PO$_4$
2.1 g Na$_2$HPO$_4$
300 mL distilled water

Mix solutions 1 and 2 and bring to a total volume of 1 L with distilled water. Autoclave.

a. Inoculate 1 mL of an overnight culture of *E. coli* W6 grown in 3×D media (Benzinger *et al.*, 1971) into 100 mL of 3×D medium. Grow with aeration at 37°C until A$_{600}$=0.3.

b. Centrifuge 60 mL of culture at room temperature at low speed to pellet the cells.

c. Resuspend the pellet in 1.0 mL of 1.5M sucrose.

PROCEDURE	COMMENTS

d. Add the following solutions in order at room temperature:

- 0.5 mL 30% bovine serum albumin-BSA (Armour sterile 30% solution or fraction V powdered BSA)
- 0.05 mL lysozyme (5 mg/mL)
- 0.12 mL of 4% EDTA, pH 8.0
- 30 mL of PA medium

e. Incubate at room temperature for 15 minutes, and add:

- 2.4 mL 10% $MgSO_4$
- 80 µL protamine sulfate (10 mg/mL)

Store at 0°C in ice.

2. Add the DNA samples (after S1 nuclease treatment step or alkaline sucrose gradient centrifugation) to 0.4 mL of 0.05M Tris, pH 8.1.

3. Incubate at 37°C for 5 minutes.

4. Add 0.2 mL of the prepared spheroplasts.

5. After 15 minutes, add 1.4 mL PAM.

6. Incubate at 37°C for 3 to 4 hours.

7. Add 0.2 mL chloroform.

8. Assay phage progeny by standard plaque assay.

d. **PA Medium:**

  100 g sucrose
    1 g glucose
   10 g casamino acids
   10 g nutrient broth (Difco)
    1 L distilled H_2O

Autoclave

e. The addition of protamine sulfate to lysozyme-EDTA treated spheroplasts of *E. coli* stimulates transfection by double-stranded DNA species. Single-stranded DNA infectivity is not stimulated.

2. For transfection, it is not necessary to remove the salts present in the reaction mixture as only 5 µL of the DNA sample is used in the transfection.

4. The prepared spheroplasts are good for at least one week when stored on ice.

5. **PAM:**

   10 g sucrose
  0.1 g glucose
  1.0 g casamino acids
    1 g nutrient broth
    2 ml 10% $MgSO_4$
  to 100 mL with distilled H_2O

7. The addition of chloroform releases phage progeny.

8. Plaque assay is carried out as described by Pagano and Hutchinson (1971).

METHOD 2: TRANSFORMATION
(M13 SYSTEM)

E. coli JM101 cells are $CaCl_2$-treated and transformed using the following protocol.

PROCEDURE	COMMENTS

1. Inoculate 5 mL supplemented M9 medium with a single colony of JM101 from a minimal plate.

2. Add 0.5 mL of overnight culture to 50 mL-supplemented M9 medium. Incubate with shaking at 37°C to an $A_{600} = 0.3$ (about 4 hours).

2. Supplemented M9 Medium

Prepare the following solutions:

a. 10× M9 Salts

    70 g $Na_2HPO_4 \cdot 2H_2O$
    30 g KH_2PO_4
     5 g NaCl
    10 g NH_4Cl
    H_2O to 1 L

b. 20% glucose solution. Filter sterilize.

c. 10 mM $CaCl_2$. Autoclave.

d. 1M $MgSO_4$. Autoclave.

e. 1% thiamine. Filter sterilize.

Add: 5 mL of Solution a (10× M9 Salts) to 45 mL distilled H_2O in a 200 mL Erlenmeyer flask. Autoclave. Cool to room temperature.

Add: 0.5 mL Solution b
        0.5 mL Solution c
         50 μL Solution d
         50 μL Solution e

3. Pellet cells in two 40 mL centrifuge tubes by centrifuging at 5000 rpm for 8 minutes.

4. Resuspend the cells in 20 mL ice cold 50 mM $CaCl_2$. Keep on ice 20 minutes.

5. Combine cells into one tube and harvest as in 3.

6. Resuspend in 5 mL cold 50 mM $CaCl_2$.

7. Pipet 0.2 mL $CaCl_2$-treated JM101 cells into culture tubes (13×100 mm). Keep on ice.

8. Add 1, 2, 5, and 10 μL neutralized (Method 2B) CC-DNA to individual tubes. Prepare three additional tubes for controls. To one tube add no DNA, to the other two add 1 and 0.1 ng double-stranded M13 vector, respectively.

8. Because the concentration of CC-DNA from the gradient is difficult to calculate, it is wise to use different amounts of CC-DNA for the transformation.

9. Incubate tubes on ice for at least 40 minutes.

10. Heat tubes in a water bath at 45°C for 2 minutes.

11. Add 2.5 mL molten YT top agar to each tube. Vortex. Pour onto a YT plate. Allow the agar to harden at room temperature for 15 minutes.

11. YT Top Agar (100 mL)

0.8 g tryptone
0.5 g yeast extract
0.5 g NaCl
0.7 g agar

PROCEDURE	COMMENTS

12. Incubate the plates at 37°C overnight. (Plaques appear in 6–9 hours).

12. The yield of phage plaques ranges from 10–200 plaques per μg of CC-DNA solution depending on the age of the CaCl₂-treated cells and the % CC-DNA obtained. The controls should result in approximately 100 and 1000 plaques, respectively. In the event that no plaques are obtained from the CC-DNA fraction, dialyze the pooled DNA against two changes of 1L of 2 mM Tris-HCl, pH 8, 0.01 mM EDTA. Transform again with the same volumes of DNA solution suggested in step 8. If this does not result in the appearance of phage plaques then concentrate the sample by extraction with 1-butanol and transform again.

F. SCREENING FOR MUTANTS

In screening for a particular mutation, four methods have been used. Method 1 is used if the mutation can be detected by sequencing with an M13 sequencing primer, another synthetic oligonucleotide, or a restriction fragment. Method 2 is used if the mutation creates or destroys a restriction site. The hybridization screen (Method 3) is the simplest and most versatile procedure, in which the labeled mutagenic oligonucleotide is used as a probe for mutant clones. Biological selection (Method 4) has also been used to screen for mutants.

METHOD 1: CHAIN TERMINATION SEQUENCING

If the only method for selection of mutants is by DNA sequence determination, possible mutants can be identified by sequencing gels run only with the single reaction corresponding to the nucleotide residue introduced into the mutant. Mutants will contain a band at this position, absent from the wild-type. A restriction fragment or other oligonucleotide downstream from the target site in the DNA may be used as a primer for DNA sequence determination by the enzymatic chain-terminator procedure.

PROCEDURE	COMMENTS

1. Use 5 μL recombinant DNA as template with only one dideoxyribonucleotide in a chain terminator reaction (single-channel sequencing). The primer can be a universal M13 primer, another synthetic oligonucleotide, or a restriction fragment. Suspected mutants are sequenced again using all four dideoxyribonucleotide mixtures (p. 173).

1. M13 universal sequencing primer is available from Bethesda Research Laboratories. Its sequence is 5'-A-G-T-C-A-C-G-A-C-G-T-T-G-T-A-3'. See DNA sequencing (Section IIIB, p. 173).

Sequence only the region where the mutation was introduced. Thus, an oligonucleotide primer (or DNA fragment) downstream of the mutation site can be used as a primer for DNA sequencing. If the mutation site is close to the M13 insertion site, the M13 primer can be used for sequencing.

METHOD 2: RESTRICTION SITE SCREENING

PROCEDURE

1. Add to a siliconized Eppendorf tube:

5 μL recombinant DNA
1 μL 10× Klenow buffer
M13 universal sequencing primer (1.5 pmol) or another synthetic oligonucleotide
H$_2$O to 10 μL

2. Mix. Heat at 55°C for 5 minutes. Cool to room temperature.

3. Add:

1 μL 0.5 mM dCTP
1 μL 0.5 mM dTTP
1 μL 0.5 mM dGTP
0.5 μL 0.1 mM dATP
5 μCi α-^{32}P-dATP
0.5 μL 10× Klenow buffer
Mix

4. Add 1 unit of DNA polymerase (Klenow fragment). Mix.

5. Incubate for 10 minutes at room temperature.

6. Add 1 μL 2.5 mM dNTPs. Mix.

7. Heat at 65°C for 10 minutes. Cool to room temperature.

8. Adjust buffer for desired restriction enzyme.

9. Add 1−5 units restriction enzyme. Mix.

10. Incubate for 1 hour at the appropriate temperature.

11. Stop by addition of 5 μL sucrose-dye-EDTA solution.

12. Load the entire sample onto a 5% nondenaturing polyacrylamide gel with 1.5 mm spacers (see Chapter 1, Section V, p. 24).

13. Electrophorese at 200 volts for 16 hours.

14. Autoradiograph as described in Chapter 1, Section XIII, p. 79.

COMMENTS

2. If a restriction fragment is used as a primer, heat the mixture for 3 minutes at 100°C, hybridize at 67°C for 30 minutes, then cool to room temperature.

11. Sucrose-Dye-EDTA Solution:

60% sucrose
0.02% (w/v) bromophenol blue
0.02% (w/v) xylene cyanol FF
0.025M EDTA

METHOD 3: HYBRIDIZATION SCREENING

<table>
<tr><td>PROCEDURE</td><td>COMMENTS</td></tr>
</table>

1. Label the mutagenic oligonucleotide using T4 polynucleotide kinase as follows:

a. Lyophilize 20 μCi γ-^{32}P-ATP (2000 Ci/mmol) in a 0.5 mL Eppendorf tube.

a. If γ-^{32}P-ATP comes in 50% EtOH, it should be lyophilized before use.

b. Add:

  20 pmol oligonucleotide
  3 μL 1M Tris-HCl, pH 8.0
  1.5 μL 0.2M MgCl$_2$
  1.5 μL 0.1M DTT
  H$_2$O to 30 μL total volume

c. Mix and add 4.5 units T4 polynucleotide kinase. Mix.

d. Incubate at 37°C for 45 minutes.

e. Heat at 65°C for 10 minutes.

e. This is done to denature the DNA.

f. Chromatograph on Sephadex G-25 to remove unincorporated γ-^{32}P-ATP.

f. See Chapter 1, Section XVII-N, p. 103, for details on chromatography.

2. Proceed with hybridization using the **DOT-BLOT METHOD**.

2. See Chapter 1, Section XI, p. 60 for a theoretical description.

a. Spot 2 μL of phage (or DNA) onto a dry sheet of nitrocellulose.

b. Air dry, then bake at 80°C for 2 hours in vacuo.

c. Prehybridize with 6× SSC plus 10× Denhardt's solution with 0.2% SDS at 67°C for one hour.

c. The prehybridization step may be carried out in a sealable plastic bag or in a closed 10 mm × 100 cm^2 plastic Petri dish.

100× Denhardt's Solution

  2% BSA
  2% polyvinyl pyrolidone
  2% Ficoll

d. Remove the prehybridization solution and rinse nitrocellulose filter in 50 mL 6× SSC for 1 minute at 23°C.

e. Place nitrocellulose in a sealable bag. Add 4 mL of probe solution for each 100 cm^2 filter. Seal bag.

e. The probe (10^6 cpm, Cerenkov) is added to 6× SCC + 10× Denhardt's solution (no SDS added).

f. Incubate at room temperature (23°C) for one hour.

g. Remove filter. Wash 3 times with 50 mL 6× SSC for 3−4 minutes each time.

h. Autoradiograph for one hour using Kodak NS-5T film at 23°C (Chapter 1, Section XIII, p. 79).

i. Preheat the wash solution to the desired temperature in a glass dish in a water bath. Wash filter at 23°C with 50 mL 6× SSC for 5 minutes.

j. Autoradiograph as in (h).

k. Repeat steps (i) and (j) using several wash temperatures.

METHOD 4: BIOLOGICAL SCREENING

Induced mutations that produce distinct biological phenotypes are conveniently recovered by appropriate selection. An example is provided by the isolation of ribosome-binding site mutants of gene *E* in φX174 (Gillam and Smith, 1983). Gene *E* is responsible for lysis of the host cells. Therefore, mutants (nonsense mutants or ribosome-binding site mutants) in gene *E* will form plaques on lawns of *E. coli* C (Su⁻) only when the medium contains bile salts and lysozyme as artificial lytic agents (Barrell *et al.*, 1976). The ribosome-binding site mutants of gene *E* in the phage φX174 were selected by initially plating the phage progeny on bile salts−lysozyme plates in *E. coli* C and then phage plaques were replica-stabbed to lawns of *E. coli* C (Su⁻) on tryptone agar plates with and without bile salts and lysozyme. Ribosome-binding site mutants should form plaques only on the plates containing bile salts and lysozyme.

IV. References

Adams, S.P., K.S. Karka, E.J. Wyles, S.B. Holder, and G.R. Galluppi. 1983. Hindered dialkylamino nucleoside phosphite reagents in the synthesis of two DNA 51-mers. *J. Am. Chem. Soc.* 105:661−663.

Astell, C.R., M.T. Doel, P.A. Jahnke, and M. Smith. 1973. Further studies on the properties of oligonucleotide cellulose columns. *Biochem.* 12:5068−5074.

Backman, K., M. Betlach, H.W. Boyer, and S. Yanofsky. 1978. Genetic and physical studies on the replication of ColEI-type plasmids. Cold Spring Harbor Symp. Quant. Biol. 43:69−76.

Barrell, B.G., G.M. Air, and C.A. Hutchison III. 1976. Overlapping genes in bacteriophage φX174. *Nature* 264:34−41.

Bechman, K., M. Ptashne, and W. Gilbert. 1976. Construction of plasmids carrying the cI gene of bacteriophage λ. *Proc. Natl. Acad. Sci. U.S.A.* 73:4174−4178.

Benzinger, R., I. Kleber, and R. Huskey. 1971. Transfection of *Escherichia coli* spheroplasts. I. General facilitation of double-stranded deoxyribonucleic acid infectivity by protamine sulfate. *J. Virol.* 7:646−650.

Boeke, J.D. 1981. One and two condon insertion mutants of bacteriophage f1. *Mol. Gen. Genet.* 181:288−291.

Borrias, W.E., I.J.C. Wilschut, J.M. Vereijken, P.J. Weisbeck, and G.A. van Arkel. 1976. Induction and isolation of mutants in a specific region of gene A of bacteriophage φX174. *Virology* 70:195−197.

Cole, C.N., T. Landers, S.P. Goff, S. Manteuil-Brutlag, and P. Berg. 1977. Physical and genetic characterization of deletion mutants of simiam virus 40 constructed in vitro. *J. Virol.* 24:277−294.

Covey, C., D. Richardson, and J. Carbon. 1976. A method for the deletion of restriction sites in bacterial plasmid deoxyribonucleic acid. *Mol. Gen. Genet.* 145:155−158.

Denhardt, D.J., D. Dressler, and D.S. Ray, eds. 1978. The single-stranded DNA phages. Cold Spring Harbor Laboratory, Cold Spring Harbor, New York.

Faye, G., D.W. Leung, K. Tatchell, B.D. Hall, and M. Smith. 1981. Deletion mapping of sequences essential for in vivo transcription of the iso-1-cytochrome c gene. *Proc. Natl. Acad. Sci. U.S.A.* 78:2258−2262.

Gillam, S., K. Waterman, and M. Smith. 1975. The base pairing specificity of cellulose-pdT$_9$. *Nucl. Acids Res.* 2:625−634.

Gillam, S., P. Jahnke, C. Astell, S. Phillips, C.A. Hutchison III, and M. Smith. 1979. Defined transversion mutations at a specific position in DNA using synthetic oligodeoxyribonucleotides as mutagens. *Nucl. Acids Res.* 6:2973−2985.

Gillam, S., and M. Smith. 1979. Site-specific mutagenesis using synthetic oligodeoxyribonucleotide primers: optimum conditions and minimum oligodeoxyribonucleotide length. *Gene* 8:81−97.

Gillam, S., and M. Smith. 1980. Use of *E. coli* polynucleotide phosphorylase for the synthesis of oligodeoxyribonucleotides of defined sequences, *In* L. Grossman, and K. Moldave (ed.), Methods in Enzymology. Nucleic Acids, vol. 65. Academic Press, New York. 687−701.

Gillam, S., C.R. Astell, and M. Smith. 1980. Site-specific mutagenesis using oligodeoxyribonucleotides: isolation of a phenotypically silent φX174 mutant, with a specific nucleotide deletion, at a very high efficiency. *Gene* 12:129−137.

Gillam, S., and M. Smith. 1983. Synthetic oligodeoxyribonucleotides for studying site-specific mutagenesis, 199−221. *In* E. Friedberg, and P.C. Hanawalt (ed.), DNA Repair, vol. 2. Marcel Dekker, Inc.

Green, C., and C. Tibbetts. 1980. Targeted deletions of sequences from closed circular DNA. *Proc. Natl. Acad. Sci. U.S.A.* 77:2455−2459.

Hanahan, D. 1983. Studies on transformation of *Escherichia coli* with plasmids. *J. Mol. Biol.* 166:557−580.

Heffron, F., P. Bedinger, J.J. Champoux, and S. Falkow. 1977. Deletions affecting the transposition of an antibiotic resistance gene. *Proc. Natl. Acad. Sci. U.S.A.* 74:702−706.

Hutchison III, C.A., S. Phillip, M. Edgell, S. Gillam, P. Jahnke, and M. Smith. 1978. Mutagenesis at a specific position in a DNA sequence. *J. Biol. Chem.* 253:6551−6560.

Itakura, K., and A. Riggs. 1980. Chemical DNA synthesis and recombinant DNA studies. *Science* 209:1401−1405.

Klenow, H. and I. Henningsen. 1970. Selective elimination of the exonuclease activity of the deoxyribonucleic acid polymerase from *Escherichia coli* B by limited proteolysis. *Proc. Natl. Acad. Sci. U.S.A.* 65:168−175.

Lai, C.J., and D. Nathans. 1974. Deletion mutants of simiam virus 40 generated by enzymatic excision of DNA segments from the viral genome. *J. Mol. Biol.* 89:179−193.

Lau, P.P., and H.D. Gray, Jr. 1979. Extracellular nucleases of *Alteromonas espejiana. Bal* 31. IV. The single strand specific deoxyribonuclease as a probe for regions of altered secondary structure in negatively and positively supercoiled closed circular DNA. *Nucl. Acids Res.* 6:331−357.

Maniatis, T. 1980. Recombinant DNA procedures in the study of eucaryotic genes. *J. Cell Biol.* 3:564−608.

Matteuci, M.D., and M.H. Caruthers. 1981. Synthesis of deoxyoligonucleotides on a polymer support. *J. Am. Chem. Soc.* 103:3185−3191.

Melgar, E., and D.A. Goldthwaite. 1968. Deoxyribonucleic acid nucleases. II. The effects of metals on the mechanism of action of deoxyribonuclease I. *J. Biol. Chem.* 243:4409−4416.

Messing, J. 1983. New M13 vectors for cloning, 20−78. *In* R. Wu, L. Grossman, and K. Moldave (ed.), Methods in Enzymology. Recombinant DNA, vol. 101. Academic Press, New York.

Miyada, C.G., X. Soberon, K. Itakura, and G. Wilcox. 1982. The use of synthetic oligodeoxyribonucleotide to produce specific deletions in the *araBAD* promoter of Escherichia coli B/r. *Gene* 17:167−177.

Mulligan, R.C., and P. Berg. 1980. Expression of a bacterial gene in mammalian cells. *Science* 209:1422−1427.

Pagano, J.S., and C.A. Hutchison III. 1971. Transfection of *E. coli* spheroplasts. *Methods Virol.* 5:79−123.

Peden, K.W.C., J.M. Pipas, S. Pearson-White, and D. Nathans. 1980. Isolation of mutants of an animal virus in bacteria. *Science* 209:1392−1396.

Pipas, J.M., S.P. Adler, K.W.C. Peden, and D. Nathans. 1980. Deletion mutants of SV40 that affect the structure of viral tumor antigens. Cold Spring Harbor Symp. Quant. Biol. 44:285−291.

Post, I.E., S. Mackem, and B. Roizman. 1981. Regulation of a gene of herpes simplex virus. *Cell* 24:555−565.

Sakonju, S., D.F. Bogenhagen, and D.D. Brown. 1980. A control region in the centre of the 5 S RNA gene directs specific initiation of transcription:1. The 5′ border of the region. *Cell* 19:13−25.

Salganik, R.L.I., G.L. Dianov, L.P. Ovchinnikova, E.W. Vorinina, E.B. Kokoza, and A.V. Mazin. 1980. Gene-directed mutagenesis in bacteriophage T7 provided by polyalkylating RNAs complementary to selected DNA sites. *Proc. Natl. Acad. Sci. U.S.A.* 77:2796−2800.

Sanger, F., S. Nicklen, and A.R. Coulson. 1977. DNA sequencing with chain-terminating inhibitors. *Proc. Natl. Acad. Sci. U.S.A.* 74:5463−5467.

Shenk, T.E., J. Carbon, and P. Berg. 1976. Construction and analysis of viable deletion mutants of simian virus 40. *J. Virol.* 18:664−671.

Shortle, D., and D. Nathans. 1978. Local mutagenesis: a method for generating viral mutants with base substitutions in preselected regions of the viral genome. *Proc. Natl. Acad. Sci. U.S.A.* 75:2170−2174.

Shortle, D., J. Pipas, S. Lazarowitz, D. Dimaio, and D. Nathans. 1979. Constructed mutants of simian virus 40, 73−92. In J.K. Setlow, and A. Hollaender (ed.), Genetic Engineering Principles and Methods. Plenum, New York.

Simons, G.F.M., G.H. Veeneman, R.N.H. Konings, J.H. van Boom, and J.G.G. Schoenmakers. 1982. Oligonucleotide-directed mutagenesis of gene IX of bacteriophage M13. *Nucl. Acids Res.* 10:821−832.

So, M., and B.J. McCarthy. 1980. Nucleotide sequence of the bacterial transposon Tn1681 encoding a heat-stable (ST) toxin and its identification in enterotoxigenic *Escherichia coli* strains. *Proc. Natl. Acad. Sci. U.S.A.* 77:4011−4015.

Tatchell, K., K.A. Nasmyth, B.O. Hall, C.A. Astell, and M. Smith. 1981. *In vitro* mutation analysis of the mating-type locus in yeast. *Cell* 27:25−35.

Temple, G.F., A.M. Dozy, K.L. Roy, and Y.W. Kan. 1982. Construction of a functional human suppressor tRNA gene in an approach to gene therapy for β-thalassaemia. *Nature* 296:537−540.

Wallace, R.B., P.F. Johnson, S. Tanaka, M. Schöld, K. Itakura, and J. Abelson. 1980. Directed deletion of a yeast transfer RNA intervening sequence. *Science* 209:1396−1400.

Wallace, R.B., M. Schöld, M.J. Johnson, P. Dembek, and K. Itakura. 1981. Oligonucleotide directed mutagenesis of the human β-globin gene: a general method for producing specific point mutations in cloned DNA. *Nucl. Acids Res.* 9:3647−3656.

Wasylyk, B., R. Derbyshire, A. Guy, D. Molko, A. Roget., R. Toule, and P.C. Chambon. 1980. Specific *in vitro* transcription of conalbumin gene is drastically decreased by single-point mutation in T-A-T-A box homology sequence. *Proc. Natl. Acad. Sci. U.S.A.* 77:7024−7028.

Weissmann, C., S. Nagate, T. Tanigudri, H. Weber, and F. Meyer. 1979. The use of site-directed mutagenesis in reversed genetics, 133−150. *In* J.K. Setlow, and A. Hollaender (ed.), Genetic Engineering Principles and Methods. Plenum, New York.

Winter, G., A.R. Fersht, A.J. Wilkinson, M. Zoller, and M. Smith. 1982. Redesigning enzyme structure by site-directed mutagenesis: tyrosyl tRNA synthetase and ATP binding. *Nature* 299:756−758.

Zoller, M., and M. Smith. 1982. Oligonucleotide-directed mutagenesis using M13-derived vectors: an efficient and general procedure for the production of point mutations in any fragment of DNA. *Nucl. Acids Res.* 10:6487−6500.

appendix 1

FORMULAE FOR BUFFERS, SOLUTIONS, AND MEDIA

A. Buffers and Solutions

10 mM Tris-HCl (pH 7.8)
100 mM NaCl
 1 mM ethylenediaminetetraacetic acid (EDTA)

10 mM Tris-HCl (pH 7.6)
500 mM KCl or NaCl

20 mM Tris-HCl (pH 8.1)
12.5 mM $MgSO_4$
12.5 mM $CaCl_2$
 0.6M NaCl
 1 mM EDTA

0.5M Tris-HCl, pH 7.5
50 mM $MgCl_2$

1.0% ficoll
1.0% polyvinyl pyrolidone
1.0% bovine serum albumin
distilled H_2O

Filter, and store at −20°C.

200 mM Tris-HCl, pH 7.5
  15 mM $MnCl_2$
1000 μg/mL bovine serum albumin

1 mg/mL DNase I
50 mM Tris-HCl (pH 7.5)
1 mM dithiothreitol
50 μg/mL bovine serum albumin
50% glycerol

  1.2M sorbitol
25 mM EDTA, pH 8.0
50 mM dithiothreitol (DTT)

The DTT should be added to the other components immediately prior to use and the solution filter sterilized.

200 mM Tris-HCl, pH 7.5
100 mM $MgCl_2$
  1 mM NAD
500 (μg/mL) bovine serum albumin

0.3M sodium acetate
1 mM EDTA
pH 7.0

0.5M ammonium acetate
0.01M $MgCl_2$
0.1 mM EDTA
0.1% SDS

EXONUCLEASE BUFFER   p. 162

70mM Tris-HCl, pH 8.0
4mM $MgCl_2$
5mM dithiothreitol

EXTRACTION BUFFER   p. 84

0.1M Tris-HCl (pH 7.6)
0.1M NaCl
1 mM EDTA

FORMAMIDE-DYE-EDTA SOLUTION   p. 173
(for denaturing gels)

0.02% bromophenol blue
0.02% xylene cyanol FF
0.025M EDTA
in 90% formamide (deionized)

5× FORWARD REACTION BUFFER   p. 75

500 mM Tris-HCl, pH 7.6
100 mM $MgCl_2$
 50 mM dithiothreitol
  1 mM spermidine

10× Hin BUFFER   p. 171

66 mM Tris-HCl, pH 7.4
66 mM $MgCl_2$
0.5 M NaCl
10 mM dithiothreitol

HOMOGENIZING BUFFER   p. 84

0.1 M Tris-HCl (pH 7.5)
 50 mM KCl or NaCl
 10 mM $MgCl_2$

KLENOW BUFFER   p. 160

50 mM Tris-HCl, pH 7.5
10 mM $MgCl_2$
0.1 mM dithiothreitol
50 mM NaCl

LIGATION BUFFER   p. 50

20 mM Tris-HCl, pH 7.6
10 mM $MgCl_2$
10 mM dithiothreitol
0.6 mM ATP

Dithiothreitol is a water-soluble reducing reagent which reduces disulfides to sulf-hydryl (SH) thus protecting the ligase enzyme from oxidation.

LYSIS BUFFER   p. 151

10 mM Tris-HCl, pH 8.0
62.5 mM EDTA
2% Triton X-100

LYTIC ENZYME BUFFER (Glusulase)   p. 138

1.2M sorbitol
5 mM dithiothreitol

Prepare and filter sterilize just prior to use.

LYTIC ENZYME BUFFER (Helicase)   p. 139

1.2M sorbitol
25 mM EDTA, pH 8.0
50 mM dithiothreitol

NICK TRANSLATION BUFFER   p. 69

0.5M Tris-HCl (pH 7.5)
0.1M $MgCl_2$
500 μg/mL bovine serum albumin (BSA)
10 mM dithiothreitol (DTT)

POLYMERASE BUFFER   p. 57

50 mM potassium phosphate (pH 7.4)
0.25 mM dithiothreitol
10 mM $MgCl_2$

PRETREATMENT BUFFER   p. 129

0.2M Tris-HCl
1.2M sorbitol
0.1M EDTA
0.1M β-mercaptoethanol, pH 9.1

RESTRICTION BUFFER   p. 36

100 mM Tris-HCl, pH 7.5
 50 mM NaCl
 10 mM $MgCl_2$
  1 mM dithiothreitol or
 10 mM β-mercaptoethanol

5× REVERSE (EXCHANGE) REACTION BUFFER   p. 76

500 mM imidazole-HCl, pH 6.6
100 mM $MgCl_2$
 50 mM dithiothreitol
  1 mM spermidine
  3 mM adenosine diphosphate (ADP)

S1 BUFFER   p. 162

50 mM sodium acetate, pH 4.5
 0.15M NaCl
 0.5 mM $ZnSO_4$

SCE BUFFER   p. 130

1.0M sorbitol
0.1M sodium citrate
60 mM EDTA, pH 5.8

SSC BUFFER (20×)   p. 61

3M NaCl
0.3M sodium citrate

20× SSPE   p. 64

20 mM Na_2EDTA
0.2M NaH_2PO_4
3.6M NaCl
pH 7.0 (Adjust with 50% NaOH)

STET BUFFER   p. 6

 8% sucrose
 5% Triton X-100
50 mM EDTA
50 mM Tris-HCl

Adjust to pH 8.0 with concentrated HCl.

STOP SOLUTION   p. 38

4M urea
50% sucrose
50 mM EDTA
0.1% bromophenol blue
pH 7.0

SUCROSE-DYE-EDTA SOLUTION   p. 182

60% sucrose
0.02% (w/v) bromophenol blue
0.02% (w/v) xylene cyanol FF
0.025M EDTA

10× TAILING BUFFER   p. 55

1M cacodylic acid, sodium salt
250 mM Tris-base (adjust to pH 7.6 with KOH)
2 mM dithiothreitol
10 mM $CoCl_2$

To avoid precipitation, 0.1M cobalt chloride is added dropwise to the buffer with constant stirring. Roychaudhury and Wu (1980) observed that Co^{++} is more efficient than Mg^+ for extending termini.

TBE BUFFER   p. 162

90 mM Tris-borate, pH 8.3
2.5 mM EDTA

TE BUFFER   p. 2

50 mM Tris-HCl, pH 8.0
20 mM EDTA

TES BUFFER   p. 8

0.03M Tris-HCl, pH 8.0
0.05M EDTA
0.005M NaCl

10× TRIS-ACETATE BUFFER   p. 20

400 mM Tris base
200 mM sodium acetate
18 mM EDTA
pH 7.8

890 mM Tris base
890 mM boric acid
 25 mM EDTA
pH 8.3

B. Media

BOTTOM AGAR   p. 108

L or LB Broth
1.5% agar

B BROTH   p. 108

10 g Bacto Tryptone
 8 g NaCl
 1 L distilled H_2O

3×D MEDIUM   p. 178

Solution 1

 3 g NH_4Cl
 0.3 g $MgSO_4$
 0.3 mL 1M $CaCl_2$
24.0 mL glycerol
15.0 g casamino acids
 3.0 ml 1% (w/v) gelatine
 600 mL distilled H_2O

Solution 2

0.9 g KH_2PO_4
2.1 g Na_2HPO_4
300 mL distilled H_2O

Mix solutions 1 and 2 and bring to a total volume of 1L with distilled water.
Autoclave.

L BROTH   p. 82

10 g Bacto tryptone
 5 g Bacto yeast extract
0.5 g NaCl
 1 L distilled H_2O

LB BROTH   p. 107

10 g Bacto tryptone
 5 g Bacto yeast extract
10 g NaCl
 1 L distilled H_2O

MINIMAL MEDIUM   p. 139

0.67% yeast nitrogen base (no amino acids)
2% dextrose
20 μg/mL of amino acids
2% agar

PA MEDIUM   p. 179

100 g sucrose
  1 g glucose
 10 g casamino acids
 10 g nutrient broth (Difco)
  1 L distilled H_2O

Autoclave

PAM   p. 179

 10 g sucrose
0.1 g glucose
1.0 g casamino acids
  1 g nutrient broth
  2 mL 10% $MgSO_4$
  to 100 mL with distilled H_2O

REGENERATION AGAR (kept at 48°C)   p. 139

1.2M sorbitol
0.67% yeast nitrogen base (no amino acids)
2% dextrose
2% agar
20 μg/mL amino acids (when necessary)

SUPPLEMENTED M9 MEDIUM   p. 180

Prepare the following solutions:
a. 10× M9 Salts

   70 g $Na_2HPO_4 \cdot 2H_2O$
   30 g KH_2PO_4
    5 g NaCl
   10 g NH_4Cl
   H_2O to 1 L

b. 20% glucose solution. Filter sterilize.

c. 10 mM CaCl$_2$. Autoclave.

d. 1M MgSo$_4$. Autoclave.

e. 1% thiamine. Filter sterilize.

Add: 5 mL of Solution a (10× M9 Salts) to 45 mL distilled H$_2$O in a 200 mL Erlenmeyer flask. Autoclave. Cool to room temperature.

Add: 0.5 mL Solution b
     0.5 mL Solution c
     50 μL Solution d
     50 μL Solution e

TOP AGAR   p. 108

L or LB Broth
0.7 to 0.8% agar

YEPD BROTH   p. 128

1% yeast extract
2% peptone
2% dextrose

Yeast cells lyse more readily following growth in media containing galactose, which can be substituted for dextrose in YEPD.

2×YT MEDIUM   p. 109

16 g Bacto tryptone
10 g Bacto yeast extract
 5 mg NaCl
 1 L distilled H$_2$O

YT TOP AGAR   p. 180
(100 mL)

0.8 g tryptone
0.5 g yeast extract
0.5 g NaCl
0.7 g agar

appendix 2

SOURCES OF REAGENTS AND EQUIPMENT

A. Apparatus[1]

1. *General*

Unless listed below, all apparatus mentioned in the manual may be purchased from either (or both):

CanLab Ltd.
80 Jutland Road
Toronto, Ontario
M8Z 2H4
Canada
(Telex: 06-219893)

Fisher Scientific
184 Railside Road
Don Mills, Ontario
M3A 1A9
Canada
(Telex: 06-966672)

Fisher Scientific
711 Forbes Avenue
Pittsburgh, PA 15219
U.S.A.

[1]These lists are not intended to be exhaustive. We apologize for any omissions.

2. *Specific*

a. Centrifuges

(micro):

Eppendorf/Brinkman (Canada) Ltd.
50 Galaxy Blvd.
Rexdale, Ontario
M9W 4Y5
Canada

Beckman Ltd.
901 Oxford Street
Toronto, Ontario
M8Z 5T2
Canada

Beckman Instrument Ltd.
117 California Avenue
Palo Alto, CA. 94304
U.S.A.

(refrigerated):

Beckman Ltd.

Sorval
Maynard Scientific Division
Ingram and Bell
22 Lido Road
Weston, Ontario
M9M 1M6
Canada
(Telex: 065-27445)

DuPont Company
Instrument Products
Biomedical Division
Newtown, CT 06470
U.S.A.
(Telex: 969632)

(ultra):

Beckman Ltd.
Sorval

b. Gel electropnoresis:

(BioRad)   **Bio-Rad Laboratories**
3140 Universal Drive
Mississauga, Ontario
L4X 2C8
(Telex: 06-961-233)

Bio-Rad Laboratories
2200 Wright Avenue
Richmond, CA 94804
U.S.A.
(Telex: 335-358)

(BRL)   **Bethesda Research Laboratories**
GIBCO Canada Inc.
2260A Industrial St.
Burlington, Ontario
L7P 1A1
(Telex: 06-18270)

Bethesda Research Laboratories
411 N. Stonestreet Avenue
Rockville, MD 20850
U.S.A.

(IBI) International Biotechnologies Inc., (Canada) Ltd. 28 Queen Street West P.O. Box 27 Toronto, Ontario M5H 3S2 (Telex: 643993)	International Biotechnologies Inc. P.O. Box 1565 New Haven, CT 06506 U.S.A.
Pharmacia (Canada) Ltd. 2044 St. Régis Blvd. Dorval, Québec H9P 1H6	Pharmacia Fine Chemicals Division of Pharmacia Inc. 800 Centennial Avenue Piscataway, NJ 08854 U.S.A.

c.   Power sources

BioRad
BRL
Pharmacia

d.   Heating blocks

Eppendorf/Brinkman

Scientific Products Division
American Hospital Supply Corp.
McGraw Park, IL 60085
U.S.A.

e.   Ultraviolet light sources

i.   hand-held
   Ultraviolet Products (Fisher, CanLab)

ii.   transilluminators
   Ultraviolet Products (Fisher, CanLab)

Fotodyne Inc.
16700 West Victor Road
P.O. Box 183
New Berlin, WI 53151
U.S.A.

f.   Vacuum oven

Fisher, CanLab

g.   Lyophilizer (Savant Speed Vac Concentrator)

Emertson Instruments 2275 Victoria Park Avenue Suite 205 Scarborough, Ontario M1R 1W4	**Savant Instruments Inc.** 221 Park Avenue Hicksville, NY 11804 U.S.A.

B. Chemicals

1. *General*

Unless listed below, all chemicals mentioned in the manual may be purchased from:

Fisher Scientific
CanLab

(BDH) **British Drug House**
350 Evan Avenue
Toronto, Ontario
M8Z 1K5
(Telex: 06-967678)

2. *Specific*

a. Antibiotics

Sigma Chemical Co.
P.O. Box 14508
St. Louis, MO 63178
U.S.A.

Calbiochem Ltd.
La Jolla, CA 92037
U.S.A.

(from antibiotic manufacturers themselves)

b. Bentonite

Sigma Chemical

c. Bovine serum albumin

BRL
Sigma Chemical
IBI

d. Cacodylic acid

Sigma Chemical

Aldrich   Aldrich
114 Rue du Fort   P.O. Box 2060
Suite 1403   Milwaukee, WI 53201
Montreal, Québec   U.S.A.
H3H 2N7

IBI

e. Cesium chloride (analytical grade)

BDH
IBI

f.   Cordycepin 5′ triphosphate

(NEN)   **New England Nuclear**
2453-46th Ave.
Lachine, Québec
H8T 3C9
(Telex: 05-821808)

Amersham
505 Iroquois Shore Road
Oakville, Ontario
L6H 2R3
(Telex: 069-82286)

New England Nuclear
549 Albany Street
Boston, MA 02118
U.S.A.
(Telex: 94-0996)

Amersham
2636 South Clearbrook Drive
Arlington Heights, IL 60005
U.S.A.

g.   Diethylbarbituric acid

Sigma Chemical

h.   Diethyl pyrocarbonate

BDH

i.   Dithioerythritol

Sigma Chemical
IBI

(BMC)   Boehringer-Mannheim Canada Ltd.
11450 Côte de Liesse Rd.
Dorval, Québec
H9P 1A9
(Telex: 05-822677)

j.   DNA (calf thymus, herring sperm, salmon sperm)

Sigma Chemical
Calbiochem
BMC

k.   Deoxyribonucleotides

Sigma Chemical
BRL
BMC
Amersham Corporation

l.   Ethidium bromide

Calbiochem
Sigma Chemical
IBI

m. Formamide

Eastman Kodak Company
Eastman Organic Company
Rochester, NY 14650
U.S.A.

Aldrich Chemical Co.
IBI

n. Imidazole-HCl

Sigma Chemical
IBI

o. Mercaptoethanol

Sigma Chemical
BMC

p. Methyl mercuric hydroxide

AISA Products
Phiokol-Ventron Div.
152 Andover St.
Danvers, MA 01923
U.S.A.

q. Nitrocefin (chromogenic cephalosporin)

(powder) **Glaxo Canada Ltd.**
1025 The Queensway
Toronto, Ontario
M8Z 5S6
(Telex: 06-967582)

(discs) **Becton Dickinson & Co., Canada Ltd.**
2464 South Sheridan Way
Mississauga, Ontario
L5J 2M8

r. Polyethylene glycol

Sigma Chemical

s. Polyvinyl sulfate

Sigma Chemical

t. Sodium sarcosinate (Sarkosyl)

Sigma Chemical
IBI

u. Spermidine

Sigma Chemical

v. Synthetic linkers

BRL
BMC
IBI

w. tRNA

BRL
BMC
IBI

x. X-gal

BRL (Blue-o-gal)
BMC
IBI

C. Chromatography Materials

1. *Membrane papers*

a. DEAE

Spectrex Ltd. (Schleicher & Schuell Papers)
5250 Ferrier St.
Suite 508
Montreal, Québec
H4P 1L6
Canada

Biorad

b. Nitrocellulose

Spectrex Ltd.
BioRad

c. DBM

BioRad

d. Nylon

BioRad ("Zeta Probe")
New England Nuclear ("Genescreen")

Pall Ltd. ("Biodyne")
1590 Malheson Blvd.
Suite 24
Mississauga, Ontario
L4W 1J1
(Telex: 6-960411)

2. *Column materials*

a. Dowex 50

BioRad

b. Sephadex

Pharmacia Ltd.

c. Oligo (dT) cellulose

BioRad
Pharmacia
BMC

d. Hydroxylapatite

BioRad
BMC

e. Acridine yellow ED gel (dye-based affinity gel)

BMC

f. NACS™ system (ion-exchanger)

BRL

g. Bio-Gel A

BioRad

h. Sepharose-CL

Pharmacia

i. Mixed Bed Resin AG501-X8, 20–50 mesh

Biorad

D. DNA and Host Strains   See Section XVII-K, Chapter 1, p. 101.

E. Enzymes

a. Bacterial alkaline phosphatase

Amersham
BRL
BMC
IBI

b. DNA polymerase I

Amersham
BRL
BMC
IBI

c. DNAase I

Amersham
BRL
BMC

d. Klenow fragment

Amersham
BRL
BMC
IBI

e. Lysozyme

Sigma Chemical
Calbiochem

f. Polynucleotide kinase

Amersham
BRL
BMC

g. Restriction endonucleases

Amersham
BRL
BMC
IBI

Miles Laboratories   Miles Laboratories, Inc.
77 Belfield Road   P.O. Box 2000
Rexdale, Ontario   Elkhart, IN 46515
M9W 1G6   U.S.A.
(Telex: 06-989399)

h. RNase A_1, T_1

BRL
BMC

i. T4 ligase

Amersham
BRL
BMC
IBI

j. T4 polymerase

Amersham
BRL
BMC
IBI

k. Terminal transferase (deoxynucleotidyl)

Amersham
BRL
BMC
IBI

F. Gelling Materials
(Agarose,
acrylamide)

BioRad
BRL
Pharmacia
IBI

Marine Colloids ("Seakem")
FMC Corporation
Bio-Products Dept.
5 Maple St.
Rockland, ME 04841
U.S.A.

G. Media
(dehydrated)

Difco Canada Inc.

Oxoid Canada Inc.
217 Colonnade Road
Nepean, Ontario
K2E 7K3
(Telex: 0533211)

Gibco Canada Inc.
2260A Industrial St.
Burlington, Ontario
L7P 1A1
(Telex: 06-18270)

H. Photographic
Supplies

a.   Camera: Polaroid MP4

Polaroid Canada Inc.
350 Carlingview Drive
Rexdale, Ontario
M9W 5G6

b.   Film: Type 52 (4×5 prints)
        Type 55 (4×5, prints, 8 negatives)

Polaroid Canada

c.   X-ray film:

Kodak: distributed by **Picker Corporation**
1750 Walkley Road
Ottawa, Ontario
K1V 8T6

d.   Developing chemicals

Kodak (distributed by Picker Corporation)

e. Exposure cassettes

Kodak, Dupont (distributed by Picker Corporation)

I. Plasticware

a. Microfuge tubes, pipette tips, Pasteur pipettes

BioRad
Beckman
Eppendorf/Brinkman

b. Counting vials (scintillation)

Fisher Scientific
CanLab

c. Poly-bags: tubing (5 cm width)

Canus Plastic Ltd.
340 Gladstone Ave.
Ottawa, Ontario
Canada
     "Seal-N-Save" (20 cm width)
Sears Ltd. (retail outlets)

J. Radioisotopes and Kits

a. $^{3}H/^{32}P$-labeled dNTPs

New England Nuclear
Amersham Corporation

b. Nick-translation kits

NEN
Amersham Corporation

c. Fluorographic spray (En^{3}Hance™)

New England Nuclear

d. M13 cloning and sequencing kits

BRL
Amersham

e.   DNA sequencing kits (Maxam-Gilbert)

NEN

Nonradioactive immunogenic probe systems are available from:

Enzo Biochem Inc.
325 Hudson St.
New York, NY 10013
U.S.A.
(Telex: ENZOB 10 424683)

BRL

INDEX

Cerenkov:
   counting, 107, 183
   light, 107
   see also Liquid scintillation counting
Cesium chloride centrifugation:
   for isolating DNA, 7, 131, 151
   for purifying phage, 110
Chaotropic agents, 35
Charon vectors:
   3A, 45
   4A, 45
      map of, 45
   16A, 45
   21A, 45
   23A, 45
   see also Bacteriophage, lambda, 28, 45
χ1776, *E. coli* strain, 48, 82
χ2098, *E. coli* strain, 48
Chloramphenicol:
   and gene amplification, 42
   resistance to, 131, 148
Chloroform, 100, 109, 178
Chromatography for isolating DNA, 9, 103
   affinity absorption, 9
   hydroxylapatite, 10
   ion exchange, 10, 103
   step elution, 11
Chromogenic cephalosporin test, for
      β-lactamase, 111
Chromosome, definition of, xvi
*Cla*I, restriction enzyme, 49, 53, 148
Clone, definition of, xvi
Cloning, *see* Gene cloning
Cloning vectors:
   autonomously replicating, 41, 132
   bacterial-yeast hybrid, 133
   bacteriophage, 44, 109, 145, 170
   cosmids, 42
   definition of, xx
   expression, 47, 136, 142
   general, 41
   integration (yeast), 132
   Pseudomonas aeruginosa, 147
   selected, table of, 43, 133
   shuttle, 47
   two micrometer DNA (yeast), 127, 132,
      134
   see also Bacteriophage; Cosmid(s);
      Plasmid(s)
Clostridium, 1, 2
Cohesive (sticky) ends of DNA molecules,
      35
   definition of, xvi
*Col*EI, 8, 26, 164
   immunity to, 101
Colony hybridization for screening for DNA
      sequence, 65, 140
   definition of, xvi
Comb, well-forming, in electrophoresis, 14,
      19
Competence for transformation, 53, 81, 178
   artificial, 81
   natural, 81
   storage of competent cells, 83
   treatment with CaCl₂, 83
Complementary (c) DNA, 57, 86

definition of, xvi
Concentrations, expression of, 97
Conjugation, definition of, xvi
Containment, to prevent environmental
      contamination of potentially
      harmful recombinant DNA
      molecules, 41
   biological, 41
      EK1, EK2, 41
      HV1, HV2, 41
      SHY strains, 138
   definition of, xxi
   physical, 41
      P1-P4, 41
Copy number of plasmids:
   high, 134
   low, 132, 133, 135
Cosmid(s), 42
   definition of, xvi
   vectors:
      pHC79, 42, 43
         map of, 44
      pKT264, 43
      pYC1, 43
Covalently-closed circular (CCC) DNA, 9,
      129. *See also* DNA, covalently-
      closed circular
   definition of, xvi, 2
CYC1 gene of yeast, 142
Cystic fibrosis patients and incidence of
      Pseudomonas infections, 147

DEAE paper for transferring DNA from
      gels, 31
Deletion mutants constructed by
      oligonucleotide-directed
      mutagenesis, 158, 167
Denaturing gels:
   preparation of, 88
   uses of, 39, 84, 173
Denhardt's reagent for prevention of probes
      from binding to nitrocellulose, 64,
      183
Dephosphorylation of DNA, 52, 75. *See also*
      Alkaline, phosphatase
Depurination of DNA by acid, 61
Destaining of gels, 28
Developing:
   of autoradiographs, 81
   of photographs, 30
   of X-ray film, 81
Dextran, for gel filtration, 103
Diazobenzyloxymethyl (DBM) paper for
      nucleic acid hybridization, 60
2,3-Dibromopropanol, for cross-linking
      agarose to form Sepharose-CL, 104
Dichlorodimethylsilane, for siliconization of
      labware, 100
   precautions for use of, 100
Dideoxy method for DNA sequencing, 90,
      173
Diethylaminoethyl (DEAE) paper for
      transferring DNA from gels, 31
Diethyloxydiformate, 128
Diethylpyrocarbonate to inhibit nuclease
      activity, 1, 3, 84

Digestion of DNA, 7, 35
2,5-Diphenyloxazole, as organic scintillator,
      80, 106
Directional cloning for orientation of cloned
      fragment, 49
Displacement loop (D-loop), formation of,
      158
Disruptive release of DNA from gels, 30
Distamycin A binding to DNA, effect of, on
      electrophoretic mobility, 37
Dithiothreitol, 36, 50, 141
DMB paper, for nucleic acid hybridization,
      60
DNA:
   amplification, 8, 40
   annealing, xv
   c- (complementary), 86
      blunt-end joining to cloning vectors, 88
      cloning of, 86
      definition of, xvi
      synthesis of, 86
   c18575am7, for molecular weight marker,
      26
   calf thymus, 64, 103
   carrier, for transfer of probes to
      nitrocellulose, 64
   centromeric (yeast), 135
   circular, 9, 32
      definition of, xvi
      methods for isolation of, 2
   concentration determination, 102
   covalently-closed circular, xvi, 2, 9, 129
   denaturation of, 3, 5, 61, 98, 174
   dephosphorylation, 52. *See also* Alkaline,
      phosphatase
   depurination by acid, 61
   end-labelling for preparing radiolabelled
      probes, 70
   foreign, in expression vector, 136
   herring sperm, 64
   heterologous, use for hybridization, 64
   hybrid, 40
   isolation of, 1, 9, 128, 151
      alkaline detergent method, 1
      boiling procedure, 6
      cesium chloride ethidium bromide
         ultracentrifugation, 7, 131, 151
      column chromatography, 9, 103
      moderate culture volumes, 4
      small culture volumes, 2
   labelling for hybridization, 67
   linear, 2, 32, 158, 159
   mapping of restriction sites, 35, 39
   methylated, 36, 39
   mitochondrial in yeast, 133
   modification, 39
      definition of, xviii
   molecular:
      nature, 25
      weight determination, 26
   nicks in, 25, 29, 61, 102
   phenol extraction of preparations, 129
   photography of, 29
   plasmid, isolation of, 1
   polymerase, xvii, 86
      Klenow fragment, 54, 165

*Taq*I, restriction enzyme, 36
TBE buffer, 162
TCA (trichloroacetic acid), for DNA
   precipitation, 108
TE buffer, 2, 192
Temperate phage, definition of, xx
Terminal transferase (terminal nucleotidyl),
   xviii, 53, 71, 73
   definition of, xx
   3'-end labelling, 71, 73
TES buffer, 8
tetracycline resistance marker, 42
Toxin ST gene of *E. coli,* 164
Tracking dyes, 21
   bromophenol blue, 21
   xylene cyanol, 93
Transcription, definition of, xx
Transduction, definition of, xx
Transfection, 81, 178
   definition of, xx
Transformation, 40, 81
   alkali cation, 140
   definition of, xx
   efficiency, 82, 138
   Enterobacteriaceae, of, 81
   M13 system, 179
   procedure, 81, 179
   Pseudomonas, 82, 154
   Saccharomyces, 129, 138
   Schizosaccharomyces, 140
Transillumination system, 30
Transitions, 167
Translation, definition of, xx
Transposition, 158
   definition of, xx
Transposon(s):
   definition of, xx
   induced mutagenesis, 45
   Tn3, 45, 148, 158
   Tn5, 45
   Tn903, 148

Transversions, 167
Trialkylmethylammonium chloride (anion
   exchanger in nucleic acid
   chromatography), 10
Tricene buffer, 69
Trichloroacetic acid (TCA), for DNA
   precipitation, 108
Tris-acetate buffer, 20
Tris-borate buffer, 20
Tris buffer, 98
Triton-X100, 78
Two-dimensional gel electrophoresis, 26
Two micrometer DNA:
   curing yeast strains of, 138
   vectors, 134
Type I, II, III restriction enzymes,
   definitions of, 35

Ultracentrifuge, conditions for running CsCl
   gradients, 7, 9
Ultraviolet light:
   effects on DNA:
      locating plasmid bands in, 11
      nicking, 25, 29, 61
      sensitivity to, 101
   germicidal lamps, 29
   incident irradiation, 29
   precautions when working with, 29
   transillumination, 30
Universal sequencing primer, 181
Urea, denaturing agent, 88

Vectors, *see* Cloning vectors; Plasmid(s)
Virus(es):
   avian myeloblastosis (AMV), 86
   bacterial, *see* Bacteriophage
   bovine papilloma, replicative function
      from, 48
   human cytomegalovirus detection in urine
      using dot-blot hybridization
      procedure, 61

RNA tumor, xix
rotavirus detection in stools using dot-blot
   procedure, 61
SV40, 48, 158

Western blot of proteins, 60
World Health Organization, 41

Xenopus (gene promoter defined by *in vitro*
   site-directed deletions), 158
*Xho*I, restriction enzyme, 45, 132, 148
*Xma*I, restriction enzyme, 148
X-ray film, use in autoradiography, 79
   developing of, 81
Xylene cyanol, tracking dye, 93

Yeast, 127
   autonomously replicating sequences
      (ARS), 133
   centromere vectors, 135
   centromeric DNA, 135
   gene expression, 136
   iso-1-cytochrome C gene (*CYC1*), 142
   mating type loci, 164
   mitochondrial DNA, 133
   phosphoglycerate kinase in, 136
   strains:
      LL20, 129, 138
      YF135, 129
   two micrometer DNA, 127, 134, 138
   meiotic segregation pattern of, 134
   vectors, *see* Plasmid(s), 131
   See also Saccharomyces, cerevisiae,
      Schizosaccharomyces pombe
YEPD, 128

0X174 and oligodeoxynucleotide-directed
   mutagenesis, 168, 174

Zymolyase, for preparation of yeast
   protoplasts, 128, 141